Disinfection of Pipelines and Storage Facilities Field Guide

by William C. Lauer and Fred J. Sanchez

American Water Works Association

Science and Technology

AWWA unites the drinking water community by developing and distributing authoritative scientific and technological knowledge. Through its members, AWWA develops industry standards for products and processes that advance public health and safety. AWWA also provides quality improvement programs for water and wastewater utilities.

Project Manager/Technical Editor: Melissa Valentine
Production: Claro Systems

Disclaimer

The authors, contributors, editors, and publisher do not assume responsi-
bility for the validity of the content or any consequences of their use. In no
event will AWWA be liable for direct, indirect, special, incidental, or con-
sequential damages arising out of the use of information presented in this
book. In particular, AWWA will not be responsible for any costs, including,
but not limited to, those incurred as a result of lost revenue. In no event
shall AWWA's liability exceed the amount paid for the purchase of this
book.

Library of Congress Cataloging-in-Publication Data

William C. Lauer.
   Disinfection of pipelines and storage facilities field guide / by William C.
   Lauer, Fred J. Sanchez.
     p. cm.
   Includes bibliographical references and index.
   ISBN 1-58321-423-2
     1. Water-pipes--Cleaning. 2. Waterworks--Cleaning. 3.
   Water--Purification--Disinfection. I. Sanchez, Fred J. II. American
Water
   Works Association. III. Title.

TD491.L38 2006
628.1'5--dc22

                            2006045997

**American Water Works
Association**

6666 West Quincy Avenue
Denver, CO 80235-3098
303.794.7711

Contents

About the Authors

William C. Lauer is senior technical services engineer for the American Water Works Association. Mr. Lauer has authored and edited more than a dozen books, and 50 articles and technical publications, covering all aspects of the drinking water industry including: water quality, treatment, reuse, distribution system operation, management, and desalination. He is a recognized technical expert in the field and has consulted for NASA, USEPA, the government of Singapore, several major engineering design and construction firms, and many others in his more than 30 years in the drinking water supply field.

Fred J. Sanchez is water quality supervisor for Denver Water. Mr. Sanchez has more than 15 years experience in water quality investigation, and field disinfection and dechlorination of pipelines and storage facilities. He has developed improved field disinfection methods and testing of water pipelines and water storage facilities including the use of ozone for disinfection purposes.

Foreword

This Disinfection of Pipelines and Storage Facilities Field Guide is one of the "Field Guide" series of books published by the American Water Works Association (AWWA). These books are meant to be small, practical, how-to publications on specific subjects of importance to drinking water system operating personnel. These books omit most of the theory and background that have led to the use of the procedures described in the guides. This fundamental information is found in other AWWA publications and is referenced in the field guides.

The information in this field guide, like the other books in this series, provides what is needed to do the work. Useful tables and easy to follow illustrations help system operators perform the procedures described in the ANSI/AWWA disinfection standards and in other references. The field guides get straight to the point and provide the necessary information to perform the *most common* procedures. This approach leads the operator to the most useful solutions.

There are several special notes included throughout the book. Look for "Ops Tips" and the "Table Tamer" in call-outs and text boxes for help with important points and how to use some of the tables. There are also several "calculators" that help plug in the numbers needed to calculate a value.

Ops Tip   Operator tips indicate important points.

Table Tamer   Illustrates how to use some of the more complicated tables.

"Calculators" provide easy-to-use, plug-in-the-numbers, equations and examples used to give quick results. The conversion factors and other constants have been combined so that the equation is greatly simplified. The derivation of the formula is not given, just the result. More detail about the source of the constants shown for the calculators is shown in Appendix A.

| Calculation result and units of measure | = | Insert known value needed for the calculations | × | constant | (cn-n) |

The authors hope this information is useful and is used to supplement hands-on training and, thus, becomes an indispensable companion for both utility and contractor personnel.

William C. Lauer
Fred J. Sanchez

Acknowledgments

The authors wish to thank the reviewers who provided the benefit of their experience to enhance this publication.

Gary Burlingame, Philadelphia Water Department, Philadelphia, Pa.

Chet Shastri, Fort Wayne City Utilities, Fort Wayne, Ind.

Stephen Lohman, Denver Water, Denver, Colo.

Kenneth Morgan, Charlotte-Mecklenburg Utility, Charlotte, N.C.

Bruce F. Dahm, Denver Water, Denver, Colo.

Nicole M. Peschel, Denver Water, Denver, Colo.

Denver Water has generously provided originals of many of the illustrations used in this book. Several of these were drawn by Nicole Peschel and Fred Sanchez. Other photos are courtesy of Denver Water.

Chapter 1

The Importance of Distribution System Disinfection

The reports of waterborne disease outbreaks attributed to drinking water have decreased since high numbers in the 1970s and 1980s. However, the annual number remains higher than desirable (Figure 1-1). Acute gastrointestinal illnesses comprise the majority of these identified disease outbreaks in the United States. Most of these outbreaks involve microbiological agents that would respond to proper disinfection.

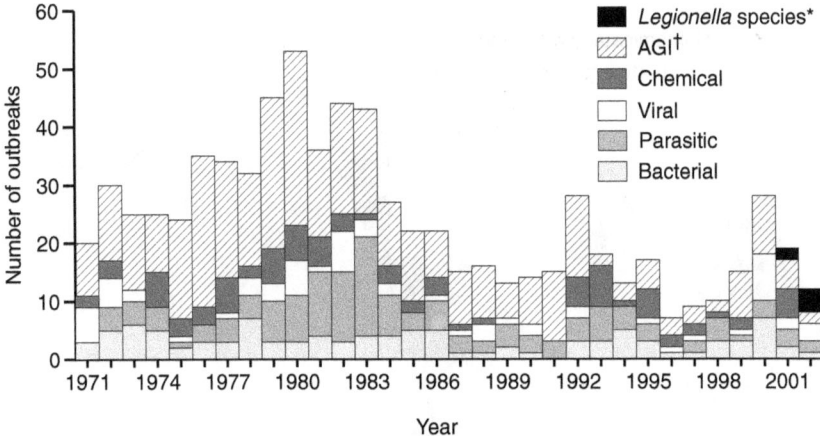

*Beginning in 2001, Legionnaire's disease was added to the surveillance system, and *Legionella* species were classified separately.
†Acute gastrointestinal illness of unknown etiology.

Reprinted from CDC MMWR Surveillance Summaries, *October 2004.*

Figure 1-1   Number of waterborne-disease outbreaks (n=764) associated with drinking water, by year and etiologic agent—United States, 1971–2002

According to the Centers for Disease Control and Prevention (CDC), deficiencies in the distribution systems are significant contributing factors leading to waterborne disease outbreaks. Figure 1-2 illustrates the relative contribution of various disease outbreak causes. Distribution system deficiencies account for 32 percent of the total, and these sources are second only to treatment deficiencies at 48 percent. Therefore, careful attention is needed to ensure the integrity of distribution systems, including disinfection of pipelines and storage facilities to reduce the possibility microbiological contamination.

Pathogens can enter the distribution system through openings in storage facilities, and during water main installation and repair procedures. Pathogens present in water or soil in proximity to water mains and storage facilities (contaminated by sewage, farmland runoff, or other polluted sources) may also enter

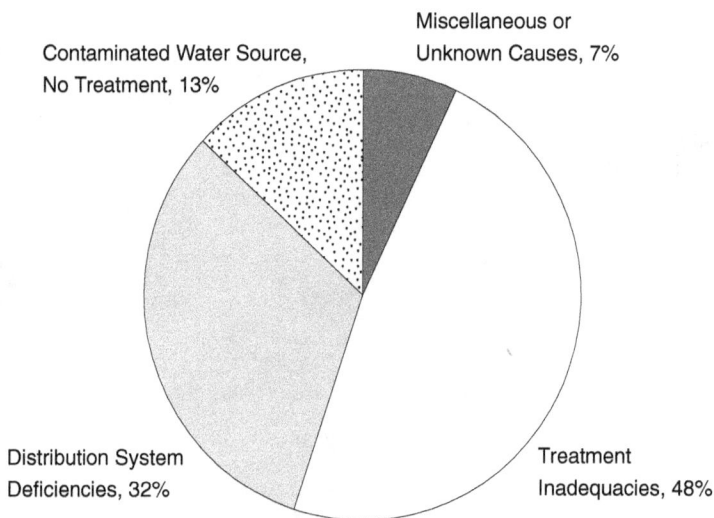

Figure 1-2   Drinking water associated waterborne disease outbreak causes, Community Water Survey, 1971–2002

pipelines or storage facilities. Figure 1-3 illustrates the findings from tests performed on undisturbed soil and water near pipe installations (Kirmeyer et al. 2001).

Soil containing pathogens and other microbes is the most common source of microbiological contamination associated with water main installation and repair. Wet soil often contains more of these contaminants than dry soil. Microbiological contamination of pipelines comes from:

- Accumulation of soil, sediment, and other foreign material on the interior of new pipes, and appurtenance surfaces during storage and installation.
- Accumulation of soil on the interior of existing exposed pipes.

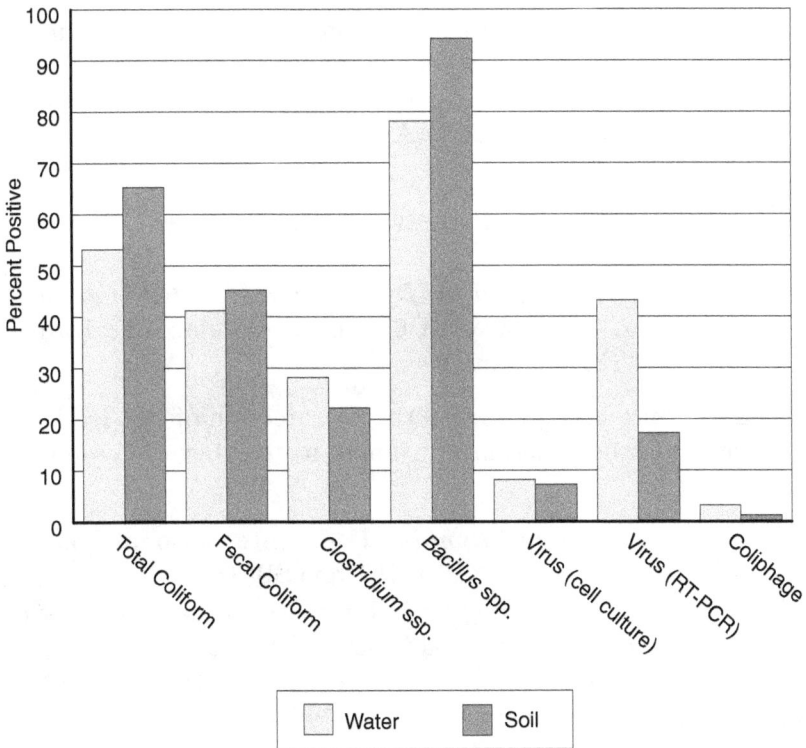

Figure 1-3   Percent positive (for various microbes) water and soil samples

- Contact with or intrusion of foreign water during storage and installation activities, and sometimes following emergency repairs, most often because of premature main shutdown and depressurization. Foreign water includes trench water, stormwater, and other sources of runoff. For main breaks, potable water that enters the trench from the broken pipe and therefore contacts contaminated soil, is a common source of foreign water.

Distribution system storage facilities consist primarily of: elevated tanks, standpipes, covered above- (or partially above-) ground reservoirs, and underground reservoirs. The sources of contamination for water storage facilities are similar to those for water mains with the addition of contact or intrusion of foreign water or matter that can occur during periodic inspections. Typical contamination routes are through missing or open vents, hatches, doors, or roof openings.

Incorporating adequate disinfection practices for water mains and storage facilities can reduce the risk from these potential contamination pathways. Disinfection requirements are described in these ANSI/AWWA standards:

> AWWA Standard for Disinfecting Water Mains. ANSI/AWWA C651. Denver, Colo.: American Water Works Association.

> AWWA Standard for Disinfection of Water-Storage Facilities. ANSI/AWWA C652. Denver, Colo.: American Water Works Association.

Plumbing codes address the disinfection of service lines, valves on service lines, and customer meters. Several prominent codes are:

> International Plumbing Code, International Code Council (formerly BOCA, ICBO, and SBCCI).

> Uniform Plumbing Code, International Association of Plumbing and Mechanical Officials (IAPMO).

> National Plumbing Code, Plumbing, Heating and Cooling Contractors Association (PHCC).

Implementing these standard requirements and using good operating practices can greatly reduce the risk of waterborne disease caused by distribution system contamination.

References

AWWA Standard for Disinfecting Water Mains. ANSI/AWWA C651. American Water Works Association: Denver, Colo.

AWWA Standard for Disinfection of Water-Storage Facilities. ANSI/AWWA C652. American Water Works Association: Denver, Colo.

Barwick, R.S., D.A. Levy, G.F. Craun, M.J. Beach, and R.L. Calderon. 2000. *Surveillance for Waterborne-Disease Outbreaks—United States, 1997–1998,* MMWR Surveillance Summaries, CDC.

Blackburn, B.G., G.F. Craun, J.S. Yoder, Y. Hill, R.L. Calderon, N. Chen, S.H. Lee, D.A. Levy, and M.J. Beach. 2004. *Surveillance for Waterborne-Disease Outbreaks Associated with Drinking Water—United States, 2001–2002,* MMWR Surveillance Summaries, CDC.

Craun, G.F., and Calderon, R.L. 2001. *Waterborne Disease Outbreaks Caused by Distribution System Deficiencies,* JAWWA September.

International Plumbing Code, International Code Council (formerly BOCA, ICBO, and SBCCI).

Kirmeyer, G., M. Friedman, K. Martel, D. Howie, M. LeChevallier, M. Abbaszadegan, M. Kamuh, J. Funk, and J. Harbour. 2001. *Pathogen Intrusion into the Distribution System.* Awwa Research Foundation: Denver, Colo.

Lee, S.H., D.A. Levy, G.F. Craun, M.J. Beach, and R.L. Calderon, 2002. *Surveillance for Waterborne-Disease Outbreaks— United States, 1999–2000,* MMWR Surveillance Summaries, CDC.

National Plumbing Code, Plumbing, Heating and Cooling Contractors Association (PHCC).

Pierson, G., K. Martel, A. Hill, G. Burlingame, and A. Godfree. 2001. *Practices to Prevent Microbiological Contamination of Water Mains.* Awwa Research Foundation: Denver, Colo.

Uniform Plumbing Code, International Association of Plumbing and Mechanical Officials (IAPMO).

Chapter 2

Chlorination Chemicals

Disinfection of water mains and storage facilities is most often performed using one of several forms of chlorine: liquid chlorine, sodium hypochlorite solution, or calcium hypochlorite granules or tablets. All chemicals used for the disinfection of water mains and storage facilities should be certified to NSF/ANSI Standard 60: Drinking Water Treatment Chemicals–Health Effects and should satisfy the requirements of applicable ANSI/AWWA standards. There are other possible disinfectants; however, this field guide is limited exclusively to a discussion of chlorination methods.

Chlorine reacts with water to form hypochlorous acid, among other products, shown in the equations below (Eq. 2-1, 2-2, 2-3). This acid can dissociate in water resulting in hypochlorite ion (Eq. 2-4). Hypochlorous acid and hypochlorite ion together are known as free chlorine residual. Hypochlorous acid is 100 times more effective as a disinfectant than hypochlorite ion. The relative amounts of these two compounds in water are largely determined by the pH (Figure 2-1). Low pH (acid) conditions favor the formation of the preferred hypochlorous acid.

Liquid chlorine reaction with water

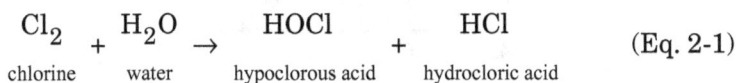

$$\underset{\text{chlorine}}{\text{Cl}_2} + \underset{\text{water}}{\text{H}_2\text{O}} \rightarrow \underset{\text{hypoclorous acid}}{\text{HOCl}} + \underset{\text{hydrocloric acid}}{\text{HCl}} \qquad \text{(Eq. 2-1)}$$

Sodium hypochlorite reaction with water

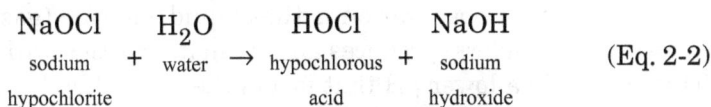

$$\underset{\substack{\text{sodium} \\ \text{hypochlorite}}}{\text{NaOCl}} + \underset{\text{water}}{\text{H}_2\text{O}} \rightarrow \underset{\substack{\text{hypochlorous} \\ \text{acid}}}{\text{HOCl}} + \underset{\substack{\text{sodium} \\ \text{hydroxide}}}{\text{NaOH}} \qquad \text{(Eq. 2-2)}$$

7

Calcium hypochlorite reaction with water

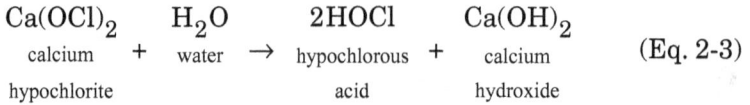

$$\underset{\substack{\text{calcium} \\ \text{hypochlorite}}}{Ca(OCl)_2} + \underset{\text{water}}{H_2O} \rightarrow \underset{\substack{\text{hypochlorous} \\ \text{acid}}}{2HOCl} + \underset{\substack{\text{calcium} \\ \text{hydroxide}}}{Ca(OH)_2} \qquad \text{(Eq. 2-3)}$$

Hypochlorous acid dissociation reaction

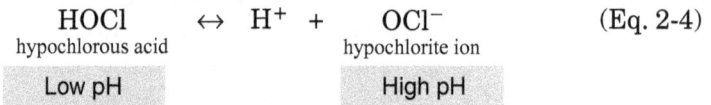

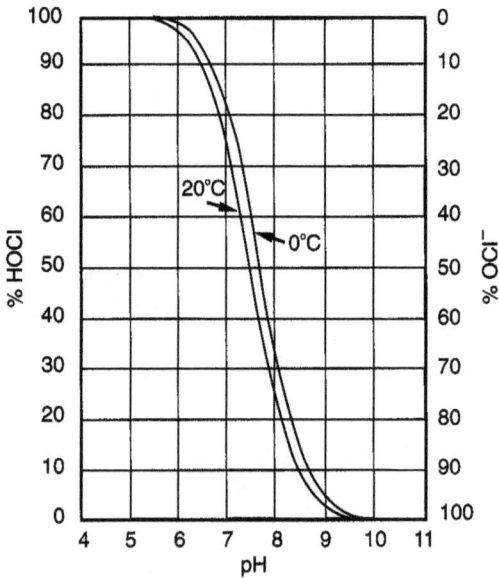

$$\underset{\text{hypochlorous acid}}{HOCl} \leftrightarrow H^+ + \underset{\text{hypochlorite ion}}{OCl^-} \qquad \text{(Eq. 2-4)}$$

| Low pH | High pH |

Figure 2-1   pH effect on chlorine species

An important observation is that liquid chlorine (gas under pressure in cylinders) produces an acid in its reaction with water. This results in a lower pH that favors the more effective form of free chlorine residual (hypochlorous acid). Both sodium and calcium hypochlorite form hydroxides in water that raise the pH. This, consequently, favors the hypochlorite ion form of the free chlorine residual. At a pH of 10, all of the free available chlorine residual is in the hypochlorite ion form. This greatly reduces the disinfection effectiveness.

> **Ops Tip** High concentrations of calcium and sodium hypochlorite raise the pH. This lowers the effectiveness of disinfection, so pH adjustment may be necessary.

The three chemicals commonly used for disinfection of pipelines and storage facilities are: liquid chlorine, sodium hypochlorite, and calcium hypochlorite. Some of the more important properties and concerns of these chemicals are discussed in the following sections.

Liquid Chlorine

Chlorine (ANSI/AWWA B301) is a liquefied gas under pressure and is normally packaged in steel cylinders, usually 100 lb, 150 lb, or 1 ton (45.4 kg, 68.0 kg, or 907.2 kg). It is also available in bulk tankers or rail cars.

At room temperature, about 80 percent of the volume of a cylinder is liquid chlorine. Liquid chlorine will change to gas very rapidly when removed from the pressurized cylinder and exposed to atmospheric conditions. One part of liquid chlorine will expand 459 times when it changes to a gas.

Chlorine is a hazardous substance. For detailed safety information, refer to publications of the Chlorine Institute, Washington D.C., and OSHA and DOT regulations that apply to chlorine handling and transportation. Safety precautions and special training are required to comply with all applicable regulations and to ensure safe handling and use. Chlorine use requires:

- Appropriate gas-flow chlorinators and ejectors to provide a controlled high-concentration solution feed.
- Direct supervision of trained personnel who are equipped to handle any emergency.

- Application of appropriate and required safety practices to protect workers and the public.
- Comprehensive knowledge, training, and certification (where required) of all applicable personnel.

Sodium Hypochlorite

This is a solution of sodium hydroxide and chlorine (NaOCl). It can be purchased in several strengths, generally from 3 to 15 percent by weight (ANSI/AWWA B300). Household bleach is 3 to 4 percent sodium hypochlorite. Commercially, it is available in a variety of containers from 1 qt (1 L) to 30 gal (113.6 L). Bulk tanks or rail cars are available in some areas. On-site generation equipment is available; therefore this chemical can be produced as needed by utility personnel or contractors.

Recommendations of the Chlorine Institute regarding safe handling specifics should be followed. Sodium hypochlorite is not as hazardous as liquid chlorine, but precautions are still recommended.

- Store in a cool, dark place to minimize the rate of decomposition.
- Store in shipping container or container made of compatible material (plastic).
- Keep container closed to prevent corrosive fumes from escaping.
- Wear personal protective equipment, such as gloves, apron, and eye protection.

Calcium Hypochlorite

Calcium hypochlorite is a dry, white or yellow-white, granular material produced from the reaction of lime and chlorine (ANSI/AWWA B300). Calcium hypochlorite is commonly 65 percent available chlorine. It is commercially available in granular, powdered, or tablet forms.

Tablet size is about 5 g and can take 7 hr or longer to dissolve in water. Colder temperatures may increase dissolving time, so adequate time is needed when using this form of the chemical. Calcium hypochlorite intended for swimming pool disinfection should not be used, because binders and other compounds may be included, making it very difficult to remove from pipelines or other facilities being disinfected.

The recommendations of the Chlorine Institute regarding safe handling specifics should be followed. Some precautions needed for the safe handling and storage of this material include:

- Store in the original container and avoid moisture.
- Decomposition may occur if stored for a lengthy period.
- It can ignite or explode on contact with organic materials (oil, rags, or alcohol) and should not be exposed to fire or elevated temperatures.
- Personal protective equipment should include: protective apron, rubber gloves, eye protection, and a dust-protection respirator.

References

ANSI/AWWA B300 *Hypochlorites*. American Water Works Association: Denver, Colo.

Chlorine Institute. 1995. *Pamphlet 100, Sodium Hypochlorite Handling and Storage Guidelines*. The Chlorine Institute: Washington D.C.

Communicating Your Chlorine Risk Management Plan. 1998. The Chlorine Institute: Washington, D.C.

Great Lakes-Upper Mississippi River Board of State Public Health and Environmental Managers. 2003. *The Ten States Standards, Recommended Standards for Water Works*. Albany, N.Y.

Molecular Chlorine: Health and Environmental Effects. 1998. Pamphlet 90. The Chlorine Institute: Washington, D.C.

Personal Protective Equipment for Chlor-Alkali Chemicals. 2001. Pamphlet 65. Chlorine Institute: Washington, D.C.

The Chlorine Manual, 6th ed. 2000. The Chlorine Institute: Washington, D.C.

US Environmental Protection Agency. *SARA Title III. Consolidated List of Chemicals Subject to Reporting Under the Emergency Planning and Community Right-to-Know Act*. USEPA: Washington, D.C.

Water and Wastewater Operators Chlorine Handbook, 1st ed. 1999. The Chlorine Institute: Washington, D.C.

White, Geo. Clifford, *Handbook of Chlorination and Alternative Disinfectants*. 1999. John Wiley & Sons: New York.

Chapter 3

Disinfection of Pipelines

Operators should follow these important steps to provide the maximum public health protection when installing new or replacement drinking water mains and other pipelines intended for potable water service. Each of these steps is discussed in more detail in the following sections.

1. Inspection
2. Sanitary Construction Methods
3. Flushing or Cleaning
4. Preventing Backflow During Installation
5. Providing Temporary Service
6. Chlorination (Chemical Disinfection)
7. Final Flushing
8. Bacteriological Testing
9. Connection to Distribution System
10. Documentation

Inspection

The pipes and appurtenances should be examined for any damage with close attention to joints. Any damage to the pipe ends or gasket areas may result in leakage. Any damaged sections should be rejected. The pipe sections should be clean and free of blemishes on the pipe interior. The coatings and linings should be fully cured before completing installation. Before placing pipe in a trench, it should be thoroughly inspected for contamination and interference.

Sanitary Construction Methods

The installation methods for water mains should include the adoption of sanitary procedures that can improve disinfection. These procedures are critical to achieve satisfactory bacteriological quality. Although it may seem time consuming or unnecessary to perform each sanitary practice, experience has shown that the steps actually save time because they reduce the failure rate of bacteriological tests. Failed tests can necessitate rechlorination or sometimes require reinstallation.

Keep Pipe Clean and Dry

The interior of pipes and pipeline fittings must be protected from contamination during installation. The openings must be sealed with watertight plugs or other suitable plugs when the construction is halted and personnel are not present to ensure that animals or debris do not enter the pipe.

> **Ops Tip**   Keeping the pipe clean during installation is the single most important factor that results in successful disinfection.

As pipe is delivered to the site and if it is strung along the trench, precautions should be taken to prevent material from entering the pipe. Any delay in the installation of delivered pipe can lead to possible contamination (Figure 3-1). One precaution is to cover the ends of the pipe if it is not installed on the delivery day (Figure 3-2). Pipe may be purchased and delivered with covered ends; this practice prevents foreign material from entering the pipe.

Joints

The joints should be connected in the trench before stopping work. Plugs should remain in place to prevent water or mud from entering the pipe.

Packing Materials

Packing material made of molded or tubular rubber rings, rope of treated paper, or other approved materials (ANSI/NSF Standard 61) should be used. Jute or hemp materials are not approved for this purpose.

Courtesy Denver Water

Figure 3-1   Unprotected pipe exposed to possible contamination

Courtesy Denver Water

Figure 3-2   Protected pipeline along trench

Sealing Materials

Only sealing materials, gaskets, and lubricants certified according to ANSI/NSF Standard 61 should be used. Lubricants should be kept in clean, closed containers and applied with clean applicator brushes.

Wet-Trench Construction

Sometimes it is not possible to prevent water from entering the pipe as it is installed. Water entering the pipe should contain approximately 25 mg/L chlorine. This can be accomplished by adding chlorine granules or tablets to each length of pipe as it is lowered into the wet trench or by treating the trench water itself.

> **Ops Tip**   Removing water from the trench will improve disinfection.

Flooding by Storm or Accident During Construction

If the new main is flooded during construction, the pipe should be drained and flushed (if possible) with potable water until the flood water is removed. The flooded section should be chlorinated so that a 25 mg/L residual remains for at least 24 hr. The chlorinated water should be drained from the main (following acceptable disposal procedures for high chlorine water). When construction is completed, the main should be disinfected using the continuous-feed or slug method described in this chapter.

Flushing and Cleaning

This may be the single most important step to ensure that the pipe passes the bacteriological tests needed before the main can be used for potable water service. Extra time and care taken with this process will result in less rework, retesting, and a quicker return to service.

> **Ops Tip**   Carefully cleaning the pipe before beginning disinfection will improve success.

Fire hydrants are usually used for flushing. In some cases, a blowoff connection is used if one has been installed. A velocity of at least 2.5 ft/sec (0.8 m/sec) is needed to remove dirt or other debris from the pipe. A low velocity flush (1 ft/sec) is adequate to "rinse" a clean pipe or to refresh the water within a main. Use Table 3-1 to determine the flushing flow rate needed to achieve 2.5 ft/sec or 1 ft/sec velocity within pipelines of various diameters.

Regardless of the velocity, flush long enough to result in two or three complete changes of the water within the pipeline. Table 3-2 is useful to determine the volume needed to achieve three

Table 3-1   Approximate pipeline flushing rates*

Pipe Diameter		Flow for 1 ft/sec velocity		Flow for 2.5 ft/sec velocity	
in.	mm	gpm	L/sec[†]	gpm	L/sec
2	50	10	0.6	25	1.6
4	100	40	2.5	100	6
6	150	1,190	6.0	200	13.0
8	200	150	10	400	25
10	250	250	16	600	38
12	300	350	22	900	57
16	400	625	39	1,600	100
18	450	800	50	2,000	126
20	500	1,000	63	2,500	158
24	600	1,400	88	3,500	220

* Flushing flow rates are approximate because precise values are not needed for these operations. Time to flush 100 ft (30.48 m) at 1 ft/sec is about 1.7 min and 40 sec at 2.5 ft/sec.
† The values in this table are calculated for US units and then converted to SI units. The L/sec values are calculated by converting the approximate gpm values. The resultant values, therefore, multiply the rounding errors.

Table Tamer   Use this table to determine the flushing rates for either a 1 ft/sec or 2.5 ft/sec velocity. Example: A 6-in. (152-mm) pipe needs to be flushed at about 200 gpm (13 L/sec) to achieve a 2.5 ft/sec velocity.

water changes within a pipeline. This volume is also the amount that will need to be dechlorinated, if that is required. A useful field estimate is to flush for about 2 minutes for each 100 ft of pipeline if your flow rate is achieving a 2.5 ft/sec velocity.

The flow may be measured by a flowmeter or estimated using the trajectory discharge method. Use Figure 3-3 to estimate the flow using this method. Table 3-3 provides some calculated values using

Table 3-2   Approximate flush volume for three water changes*

Pipe Diameter		Volume in 100 ft of pipe		Volume for Three Water Changes in 100 ft of pipe	
in.	mm	gal	L	gal	L
2	50	16	62	11,49	,185
4	100	65	247	1,196	,740
6	150	148	555	1,440	1,660
8	200	261	988	,780	2,960
10	250	408	1,540	1,220	4,630
12	300	587	2,220	1,760	6,660
16	400	1,040	3,950	3,130	11,850
18	450	1,320	5,000	3,960	15,000
20	500	1,630	6,170	4,890	18,520
24	600	2,350	8,800	7,040	26,660

* Time to flush three water changes of 100 ft (30.48 m) at 1 ft/sec is about 5 min and 2 min at 2.5 ft/sec. The values in this table are calculated for US units and then converted to SI units. Approximate volumes are acceptable for field applications.

Table Tamer   This information is used to calculate the volume of water within a pipeline and the amount needed for three water changes. Example: 500 ft of 6-in. (152-mm) pipe needs 440 gal for 100 ft (from the table), so for 500 ft, 2,200 gallons are needed.

Ops Tip   For each 100 ft (30.5 m) of pipe length flush for about 2 min.

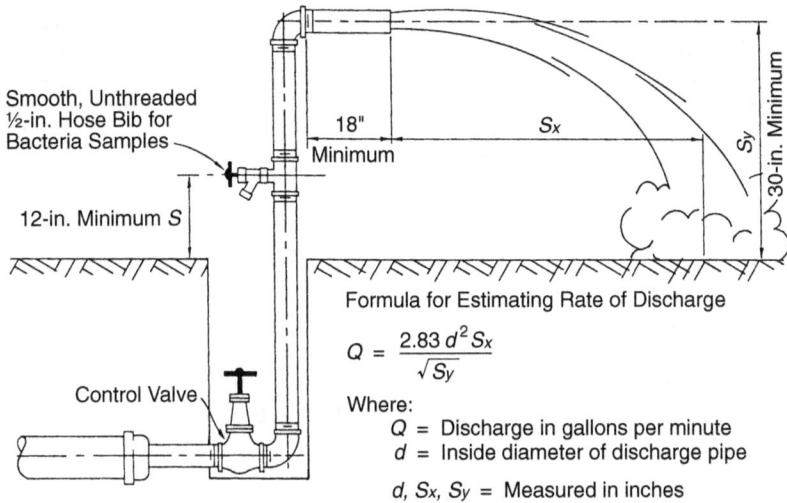

Smooth, Unthreaded ½-in. Hose Bib for Bacteria Samples

18" Minimum

S_x

S_y

30-in. Minimum

12-in. Minimum S

Control Valve

Formula for Estimating Rate of Discharge

$$Q = \frac{2.83\, d^2 S_x}{\sqrt{S_y}}$$

Where:

Q = Discharge in gallons per minute

d = Inside diameter of discharge pipe

d, S_x, S_y = Measured in inches

Note: This figure applies to pipes up to and including 8-in. (200-mm) diameter.

Figure 3-3   Trajectory discharge method of flow estimation and illustration of combination blowoff and sampling tap

this method for common size pipelines and other standard conditions. Using this table, a flow can be easily estimated from field measurements.

Large diameter pipelines may require excessive water to flush adequately. In these cases, it may be necessary to employ poly pigs or swabs (powerwashers and hydraulic jet sprayers are sometimes used) to clean the pipe without using great volumes of water (Figures 3-4, 3-5, 3-6).

If dirt enters the pipe, it should be removed by flushing the pipe with potable water, and the interior surface of the pipe swabbed with 1 to 5 percent chlorine solution (sodium or calcium hypochlorite). Small quantities of 1 percent solution should be prepared as shown in Table 3-4. If there is material adhering to the interior of the pipe that cannot be flushed out, a more aggressive cleaning may be necessary. Hydraulically propelled swabs, poly pigs, or other suitable devices including mechanical scrapers may be used to remove the attached material. This cleaning

Table 3-3   Trajectory needed to achieve flushing velocity

Diameter of Main		Diameter, d, of Discharge		Trajectory (Sx) for 1 ft/sec		Trajectory (Sx) for 2.5 ft/sec	
in.	mm	in.	mm	ft	m	ft	m
6	152	2	51	4	1.2	8.8	2.7
6	152	2.5	64	2.5	0.8	5.7	1.7
6	152	4	102	1	0.3	2.2	0.7
8	203	2	51	6.6	2	17.7	5.4
8	203	2.5	64	4.2	1.3	11.3	3.5
8	203	3	76	3	0.9	7.9	2.4
8	203	4	102	1.7	0.5	4.4	1.3

NOTE: 36 in. (914 mm) distance from discharge to ground (Sy in Figure 3-3)

Table Tamer   This table lists some typical results when using the trajectory discharge method and the equation shown in Figure 3-3. The table assumes that the distance from discharge to the ground (Sy) is 36 in. The calculation is provided so that *only* the trajectory distance from the discharge pipe is needed. Example: A 6-in. main is flushed using a 2.5-in. diameter discharge pipe. To achieve a 2.5 ft/sec velocity in the main, the trajectory must be 5.7 ft.

should be followed with an application of chlorine solution as previously described.

More aggressive methods of cleaning larger diameter pipelines include using powerwashers or hydraulic spray jets. The equipment, used commonly for cleaning sewer collection system piping, is suitable for this purpose (Figure 3-7, 3-8). These devices may be self-propelled and can do a superior job cleaning the pipe interior. Some high pressure powerwashers can clean an entire pipe section from one end.

Preventing Backflow During Installation

High chlorine levels used for disinfection of new pipelines or other nonpotable conditions may contaminate the potable

Courtesy Denver Water

Figure 3-4 Powerwashing pipe section

Courtesy Denver Water

Figure 3-5 Powerwashing close-up

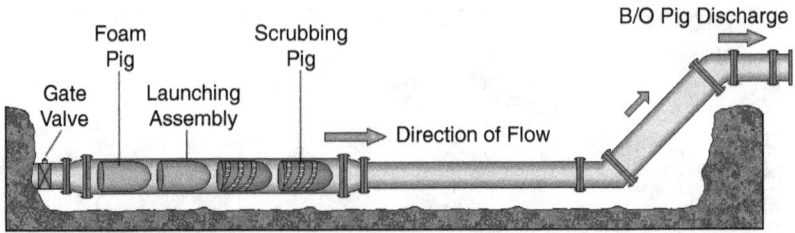

Courtesy Nicole Peschel

Figure 3-6   Pipeline pigging illustration

Table 3-4   Making small amounts of 1% or 5% chlorine solution with 65% available chlorine calcium hypochlorite*

US Units			SI Units		
Water Volume (*gal*)	1% Solution (*lb* needed)	5% Solution (*lb* needed)	Water Volume (*L*)	1% Solution (*g* needed)	5% Solution (*g* needed)
0.25 or 1 qt	0.03125 or about 3 teaspoons	0.15625	1	15.4	77
1	0.125 or ⅛ lb	0.625	4	61.6	308
2	0.25 or ¼ lb	1.25	5	77	385
3	0.375 or about ⅓ lb	1.875	8	123.2	616
4	0.5 or ½ lb	2.5	10	154	770
5	0.625	3.125	20	308	1,540
10	1.25	6.25	40	616	3,080

* The values in this table are rounded and approximated for field use.

Table Tamer   For one gallon of 1% chlorine solution, ⅛ lb or about 60 g of calcium hypochlorite should be used. See shaded cells above.

Courtesy Denver Water
Figure 3-7   Hydraulic jet cleaning

Courtesy Denver Water
Figure 3-8   Washing during pipeline repair

distribution system. Therefore, prevent backflow from newly installed (but untested) pipelines to the active distribution system. This can be accomplished with a physical separation (Figure 3-9) between the systems or by the use of an approved backflow prevention device. Hydrostatic pressure testing of new pipelines is another situation when backflow is a possibility.

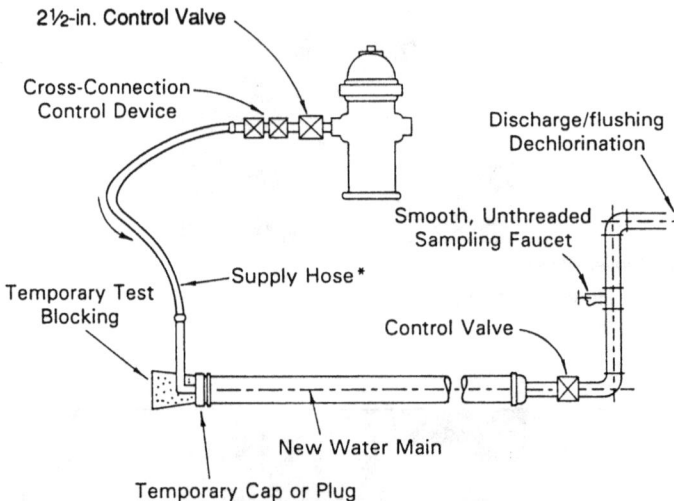

Figure 3-9   Suggested temporary flushing connection (AWWA Standard 651 *Disinfecting Water Mains*)

Providing Temporary Service

When new pipelines are installed as replacements, interruptions in water service may occur. It is common practice in the US to install temporary piping to provide continuous service when the supply is disrupted for more than several hours. Temporary supply lines are typically installed aboveground (Figure 3-10), and residences and businesses are connected by flexible hose to outside spigots or in meter pits.

The use of temporary water lines can significantly increase the project time and cost. This is because the temporary network must be disinfected (Figure 3-11), and the water must pass bacteriological testing before customers can be connected. Additionally, the public can view the temporary lines as an unsightly

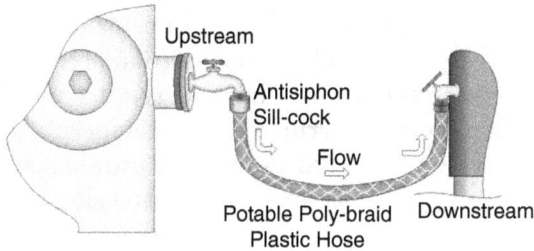

Courtesy Nicole Peschel
Figure 3-10   Temporary hydrant connection

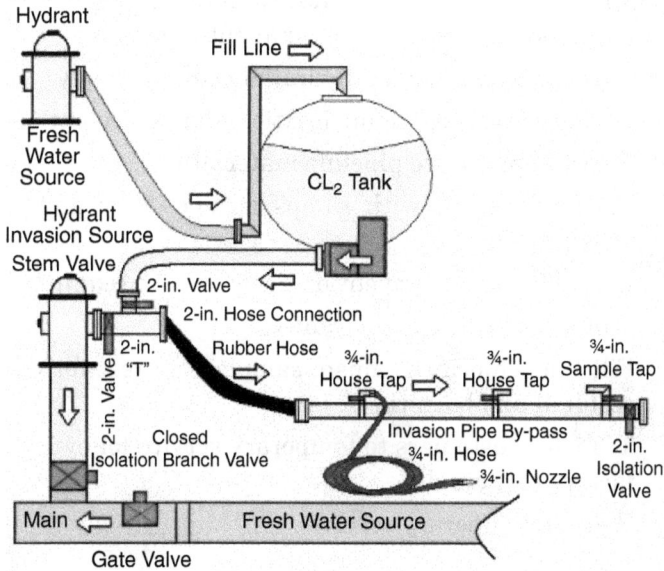

Courtesy Nicole Peschel
Figure 3-11   Temporary service line chlorination example

nuisance that may impede vehicular traffic and cause hazards for pedestrians.

Many other industrialized countries do not provide temporary service lines. Instead they select alternatives such as supplying bottled water or locating water trucks in the area. Some suppliers consider lengthy interruptions in water service as acceptable. A decision to install temporary service lines or other alternative water sources requires careful consideration of a number of factors including: site conditions, fire flow requirements, expected duration of the service interruption, cold temperatures, backflow prevention devices, internal plumbing, number of customers affected, and the type of customer (e.g., hospitals, schools, or individual residences).

Temporary pipelines are made of a variety of materials, including steel, polyvinyl chloride (PVC), and high density polyethylene (HDPE). The use of fire hoses or rubber hoses should be avoided. Components must be certified to NSF Standard 61 and approved for potable water service. Generally, installation follows the same procedures as for permanent pipelines.

- Flush and disinfect all hydrants used as supply
- Carefully select pipeline location and placement
- Select appropriate pipeline material
- Install pipeline, with ramps as necessary and buried where required
- Install backflow prevention device where required
- Connect pipeline to hydrant
- Fill line, chlorinate, flush, sample, and test for chlorine residual and bacteria
- Transfer customers to temporary service following satisfactory test results
- Remove temporary service

Ops Tip   Store short, predisinfected coils of approved pipe so they are ready for emergency use.

Disinfection procedures are the same as for permanent pipelines (ANSI/AWWA C651). Temporary pipelines must be flushed (usually at 2.5 ft/sec) to remove any dirt or debris. Chlorination is conducted according to C651 and as described in the following section. Bacteriological testing is then conducted before the temporary service is placed into use.

Chlorination

Three methods of chlorination are described in this field guide: tablet or granules, continuous feed, and slug. Factors to consider when selecting one of these methods are the length and diameter of the main, the type of joints, availability of materials, equipment required, training of the personnel, and safety concerns.

Each method has advantages and disadvantages that need to be considered. The calculation of the amount and the application procedures for each method are discussed in detail later in this chapter (summary shown in Table 3-5).

Tablets (granules)

- Pipe must be clean and dry
- Convenient for use in mains up to 24 in. diameter
- Need adequate time to dissolve in water
- Water flow may move granules to one end of a pipeline
- Disinfection may not be uniform

Continuous feed

- Suitable for general application
- Gas chlorinator needed
- Chlorine gas safety issues

Slug

- Suitable for large-diameter mains
- Chlorine gas feed and safety issues
- May save chemicals
- Reduces volume of high-chlorine water for disposal

When employing any of the chlorination methods, there are several measurements that are always needed. Some of these are defined by the pipeline measurements and some are the result of the chlorination method selected. Filling out the chlorination chemical amount calculation checklist ensures that the procedure

Table 3-5   Pipeline disinfection methods summary

Method	Chlorine Dosage	Holding Time	Minimum Chlorine Residual	Notes
Tablet or granules	25 mg/L	24 hr, 48 hr if temperature <5°C	Detectable	Add enough to fill entire pipeline segment
Continuous feed	25 mg/L	24 hr	10 mg/L	Add enough to fill entire pipeline segment
Slug	100 mg/L	3 hr	50 mg/L at any point	Fill % of pipeline with slug and move the slug slowly to give 3 hr contact

is completed satisfactorily. This checklist is used for all of the examples shown in this field guide.

Chlorination Chemical Amount Calculation Checklist
❏ The pipeline diameter and length.
❏ Valves, hydrants, tees, and other appurtenances.
❏ Volume of chlorination (Table 3-6) water to disinfect pipeline (this is the volume for disposal; chlorine residual is needed to determine dechlorination chemical amount needed).
❏ Amount of chlorine chemical needed to complete the disinfection procedure.

Final Flushing

Heavily chlorinated water should be flushed from the main following the required holding period (usually at least 24 hr). All main fittings, valves, and branches must be flushed. Chlorine

Ops Tip

This table may be one of the most useful. Just multiply the length of pipe by the volume per foot. This equals the total volume.

Table 3-6   Approximate pipeline volume for various diameters

US Units		SI Units	
Pipe Diameter		Pipe Diameter	
in.	*gal/ft*	*mm*	*L/m*
2	0.16	50	1.96
4	0.65	100	7.85
6	1.47	150	17.66
8	2.61	200	31.40
10	4.08	250	49.06
12	5.87	300	70.65
16	10.44	400	125.60
18	13.21	450	158.96
20	16.31	500	196.25
24	23.49	600	282.60
30	36.70	800	502.40
36	52.85	900	635.85
42	71.93	1,000	785.00
48	93.95	1,200	1,130.40
54	118.90	1,400	1,538.60
60	146.80	1,500	1,766.25
64	167.02	1,600	2,009.60

Table Tamer

This is two tables. The left table is in US units and the right table is in SI units. Multiply the length of the pipeline by the volume for each unit length to get the total volume of the pipeline. Example: 100 m of 150-mm pipeline contains 1,766 L.

residual of the flush water exiting the chlorinated main should be measured until the residual is no greater than the feed water.

Flush water containing high chlorine concentrations may require treatment to avoid adverse environmental impact. Consult appropriate regulatory agencies regarding requirements for the disposal of heavily chlorinated water (Tikkanen, 2001). It may be necessary to dechlorinate the water prior to disposal. Chapter 7 includes procedures for dechlorination. Some of these procedures may not be approved by every regulatory agency. Follow regulatory agency directives regarding the approved treatment procedures for your area.

Bacteriological Testing

Perform bacteriological tests on water in the new pipeline after flushing but before it is connected to the distribution system. The tests must satisfy all applicable regulatory requirements. A certified laboratory must perform the analyses. The results must be negative (no coliform present) before connecting the pipeline and releasing the water for use by customers. The following testing protocol is listed in ANSI/AWWA Standard C651.

1. Collect two samples from the new pipeline taken at least 24 hr apart.
2. Collect a sample from at least every 1,200 ft (366 m) of new main.
3. Collect a sample from the end of the line and at least one from each branch.
4. Test samples for total coliform bacteria in accordance with *Standard Methods for the Examination of Water and Wastewater*. Additional tests may be required including: chlorine residual, turbidity, pH, and heterotrophic plate count (HPC).
5. *Special Condition.* Collect additional samples at intervals of approximately 200 ft (61 m) if trench water has entered the new main during construction or if excessive quantities of dirt or debris have entered the new main. Take samples of water that has stood in the new main for at least 16 hr after final flushing.

6. Collect samples in sterile bottles treated with sodium thiosulfate as required by *Standard Methods for the Examination of Water and Wastewater*. Do not use hose or fire hydrants for the collection of samples. A combination blowoff and sampling tap used for 8 in. (200 mm) mains and smaller is shown in Figure 3-3. Ensure that there is no water in the trench up to the connection for sampling. Use a clean and disinfected sampling pipe that has been flushed prior to sampling. A corporation cock may be installed in the main with a copper tube gooseneck assembly. This assembly can be removed after sampling and used again.

7. If the HPC test results are greater than 500 colony-forming units (cfu) per mL, flush and collect a repeat sample until no coliforms are present and HPC is below 500 cfu/mL.

8. Coliform bacteria must be absent from the samples and the bacteriological quality of the water equal to or better than that of the distribution system.

9. If unsatisfactory test results are obtained, flush the main again and resample. If check samples also fail to produce acceptable results, rechlorinate the main by the continuous-feed or slug method until two consecutive sets of acceptable tests are taken at least 24 hr apart. In some cases, it may be necessary to pig or pressure wash the pipe prior to rechlorinating the main. It is advisable to check the quality of the water entering the new main because high velocities used for flushing may have disturbed sediment in the supply piping and resulted in poor quality feed water.

Connection to Distribution System

When all bacteriological tests are satisfactory the new main may be connected to the distribution system and placed in service. If the final connection is one pipe length (18 ft or 5.5 m) or less, the new pipe, fittings, and valve(s) required for the final connection may be spray-disinfected or swabbed with a 1–5 percent chlorine solution (Table 3-4) just prior to being installed. If the final connection is greater than one pipe length (greater than 18 ft or 5.5 m), the connection pipe is disinfected and bacteriological samples tested as previously described. The pipe should be connected to the distribution system after obtaining satisfactory test results. The ends of the predisinfected connection pipe should be covered during the testing and before the connection is complete.

Note: The bacteriological test is the final step in a sanitary installation process that includes inspection, use of proper materials and protection of pipes during installation, cleaning and flushing, and employing adequate disinfection procedures.

Documentation

Accurate records should be kept to document the installation conditions and test results (including water quality testing). The records should identify persons that conducted tests and that released the pipeline for service. The as-built plans should be completed identifying the types and locations of valve boxes, hydrants, and other appurtenances. The valve opening direction and any flow test results for hydrants should be recorded.

Pipeline Chlorination Method Specifics

Specific procedures for implementing each of the three pipeline chlorination methods are described in the following section. The procedures include tables and calculators for determining the necessary information to efficiently and accurately chlorinate pipelines.

Tablet Method

Calcium hypochlorite tablets (or granules) are placed in the water main as it is being installed. The main is then filled with potable water. This procedure cannot be used on solvent-welded plastic or on screwed-joint steel pipe. Fire or explosion may result from the reaction of calcium hypochlorite and joint compounds. This method also cannot be used if the main must be flushed before it is disinfected, because this process would remove the calcium hypochlorite from the main.

Use Tables 3-7a & b and 3-8a & b to determine the amount of calcium hypochlorite granules or the number of tablets needed to disinfect the length of pipeline. Table 3-7a lists the amount (by weight) of calcium hypochlorite granules needed to achieve a 25 mg/L dosage for each 500 ft of pipeline. Weights are given in ounces and grams.

Often facilities are not available to weigh out the amount of chemical needed for the job. Kitchen measures are then handy for this purpose. Teaspoons, ¼-cup, ½-cup, and 1-cup measures are readily available and Table 3-7b lists the approximate number of these measures needed. From the table, about 1 cup is needed for each 500-ft length of 8-in. diameter pipe. A plastic one-cup kitchen measure is all you need in the field to make this measurement.

Tables 3-8a & b list the number of 5-gram tablets needed for various pipe lengths, in both US and SI units. So, from Table 3-8a, a single tablet is needed for an 18-ft length of 6-in. diameter pipe. Similarly, from Table 3-8b, two tablets are required for a 6-m length of 200-mm diameter pipe.

If this method is used, great care must be taken to ensure that the pipeline is clean, dry, and free of dirt, debris, and other contamination. Fill the pipeline slowly to ensure that the granules are not moved to one end of the pipeline and to allow time for the tablets to dissolve.

Chlorination Steps and Precautions.

1. Place calcium hypochlorite granules or tablets at the upstream end of the first section of pipe, at the end of each branch main, and at 500-ft intervals (Tables 3-7a, b and 3-8a, b).
2. Use NSF/ANSI Standard 61 certified adhesive to attach tablets inside and at the top of each section of the main (Figure 3-12).

Ops Tip
Tablets aren't generally used if more than five need to be used in one place. Use a combination of tablets and granules in this case.

3. Place an equal number of tablets at each end of a given pipe length. Place one tablet in each hydrant, hydrant branch, and other appurtenance.

4. Gently fill the main with water. Ensure that the filling rate does not result in a velocity greater than 1 ft/sec (0.3 m/sec). Eliminate air pockets.

Ops Tip   Fill the line slowly (1 ft/sec or less) to avoid flushing all of the granules to one end of the pipe. See Table 3-1.

Table 3-7a   Calcium hypochlorite granules for 25 mg/L dosage by weight for each 500-ft (152.4-m) interval

Pipe Diameter		Calcium Hypochlorite Granules	
in.	mm	oz	g
2	50	0.4	12
4	100	1.7	48
6	150	3.8	1,107
8	200	6.7	190
10	250	10.5	297
12	300	15.1	428
14	350	20.4	582
16	400	25.1	761
18	450	31.8	963
20	500	39.1	1,189
24	600	56.5	1,712

Table Tamer   Table 3-7a is used to determine the weight of calcium hypochlorite needed to achieve a 25 mg/L chlorine residual in a 500-ft length of pipe of different diameters. Example: Amount needed for 500 ft of 6-in. pipe = 3.8 oz or 107 g (about ¼ lb). See shaded row above.

Table 3-7b   Handy field measures for calcium hypochlorite granules (approximately 25 mg/L dosage for each 500-ft (152.4-m) interval)

Pipe Diameter		Approximate Field Volume Measure				
in.	mm	Grams	Teaspoons* (5 mL)	¼ cup† (60 mL)	½ cup‡ (125 mL)	cups§ (250 mL)
2	50	12	2			
4	100	48	8	1		
6	150	113	19	2	1	
8	200	200	34	3	2	1
10	250	300	50	4	2	1
12	300	430	72	6	3	2
14	350	616	103	8	4	2
16	400	759	127	10	5	3
18	450	963	161	13	6	3
20	500	1,184	198	16	8	4
24	600	1,712	286	23	11	6

* A level teaspoon is about 6 g of calcium hypochlorite granules. One teaspoon is about 5 mL.
† A ¼-cup measure is about 75 g of calcium hypochlorite granules. A ¼ cup is about 60 mL.
‡ A ½-cup measure is about 150 g of calcium hypochlorite granules. A ½ cup is about 125 mL.
§ A cup measure is about 300 g of calcium hypochlorite granules. A cup is about 250 mL.

Table Tamer   The number of measures of each size. Example: For each 500-ft length of 8-in. pipe you need three ¼-cup measures of calcium hypochlorite granules. See shaded cell above.

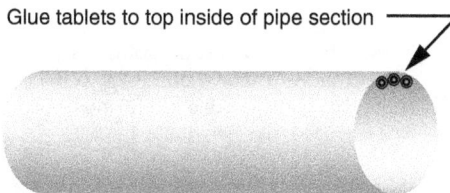

Glue tablets to top inside of pipe section

Figure 3-12   Placement of tablets inside pipe section

Table 3-8a   Calcium hypochlorite tablets for 25 mg/L dosage—US unit pipe lengths

Pipe Diameter	Length of Pipe Section, *ft*					
in.	13 or less	18	20	30	40	48
4	1*	1	1	1	1	1
6	1	1	1	2	2	3
8	1	2	2	3	4	4
10	2	3	3	4	5	6
12	3	4	4	6	7	9
16	4	6	7	10	13	
18	6	7	8	12	–	–
20	7	9	10	–	–	–
24	9	13	14	–	–	–

* Number of 5-g calcium hypochlorite tablets for each length of pipe section shown.

Table 3-8b   Calcium hypochlorite tablets for 25 mg/L dosage—SI unit pipe lengths

Pipe Diameter	Length of Pipe Section, *m*					
mm	4	5	5.5	6	10	12	16
100	1*	1	1	1	1	1	1
150	1	1	1	1	2	2	3
200	1	2	2	2	3	3	4
250	2	2	2	3	4	5	6
300	3	3	3	4	6	7	9
350	3	4	4	5	7	9	12
400	4	5	5	6	10	11	–
450	5	6	7	7	12	–	–
500	6	8	8	9	–	–	–
550	7	9	10	11	–	–	–
600	9	11	12	13	–	–	–

* Number of 5-g calcium hypochlorite tablets for each length of pipe section shown.

Table Tamer   These tables show the approximate number of tablets needed for the diameter and length of the pipe listed in the table. Example: To achieve a 25 mg/L dosage, a 300-mm pipe 6 m long will need about 4 tablets.

5. Contain the water in the pipe for at least 24 hr. Hold the water for at least 48 hr if the water temperature is less that 41°F (5°C). Optionally, water may be supplied from a temporary connection with an appropriate backflow prevention device. Bleeding a small amount of water from the pipe occasionally during the holding period will ensure that chlorinated water is well distributed throughout the pipe.

6. A detectable chlorine residual must be present at each sampling point (see section on bacteriological testing) after the holding period.

Chlorine Amount Calculation

Use Tables 3-7a & b and 3-8a & b to determine the amount of chlorine needed using the tablet method for pipeline disinfection. This method requires that chlorine in the form of tablets or granules is placed in the pipe so that when filled with water the chlorine dosage is at least 25 mg/L.

Use the chlorination chemical amount calculation checklist on page 28.

❑ The pipeline diameter and length are used to calculate the volume of the pipeline from Table 3-6.

❑ The number of valves, hydrants, and other appurtenances determine the number of tablets needed for disinfection of these items. Usually one tablet for each is adequate.

❑ This method requires that the entire pipeline is filled with water so the volume of the pipeline is the same as the volume of chlorination water (Table 3-6)

❑ The amount of chlorine chemical needed for the job comes from Tables 3-7a, 3-7b, 3-8a, or 3-8b, depending on the length and size of pipe and whether granules or tablets are used. Table 3-7b shows the approximate number of kitchen measures needed if there is no way to weigh the chemical in the field.

Study the examples below to gain a more complete understanding of how to make this calculation.

Example 1.   A 6-in. diameter pipeline consisting of 10 sections of 18-ft pipe is to be chlorinated using chlorine tablets and the tablet method. Calculate the number of tablets needed for the job and describe how they will be placed. There is one fire hydrant installed along the pipeline.

Use the checklist to ensure all of the information is collected.

❑ The pipeline diameter and length.

6-in. diameter and 10 sections of 18 ft or 180 ft total length

❑ Valves, hydrants, tees, and other appurtenances.

One hydrant along the pipeline. Disinfect with 1 tablet.

❑ Volume of chlorination water to disinfect pipeline.

Tablet method does not use chlorinated water to fill the pipeline, but the volume of the pipeline from Table 3-6 is 1.47 gal/ft × 180 ft = 264.60 gal. Fill the line slowly: from Table 3-1, a 90-gpm fill rate will be 1 ft/sec velocity. Filling the line at this rate will take about 3 min.

❑ Amount of chlorine needed to complete the disinfection procedure.

From Table 3-8a, each 18-ft section requires 1 tablet, so 10 tablets are needed for the job. Place 5 tablets at the front end of the first section and 5 at the beginning of the 6th section.

Example 2.   The tablet method is used to chlorinate a 200-mm diameter pipeline that is 1,000 m in length. Calcium hypochlorite granules are used and must be measured in the field without a scale. Determine how much is needed for the job and describe the procedure.

Use the checklist to ensure all of the information is collected.

❑ The pipeline diameter and length.

200-mm diameter and 1,000-m length

- ❏ Valves, hydrants, tees, and other appurtenances.
  No valves or hydrants are listed.
- ❏ Volume of chlorination water to disinfect pipeline.
  The tablet method does not use chlorinated water to fill the line. The pipeline volume from Table 3-6 is 31.40 L/m ×1,000 m = 31,400 L. Fill the pipeline no more than 1 ft/sec fill rate. For 200-mm pipe that is 9.4 L/s. At that rate it will take 31,400/9.4 = 3,340 sec or 56 min to fill the line.
- ❏ Amount of chlorine needed to complete the disinfection procedure.
  From Table 3-7b, one 250-mL measure of calcium hypochlorite granules is needed for each 152-m length of 200-mm pipe length. Therefore, 1,000÷152 = 6.6. So, about seven 250-mL measures (1,750 mL) of calcium hypochlorite granules are needed. Place one 250-mL measure of granules about every 150 m.

Continuous-Feed Method

In this method water from the distribution system flows slowly into the new pipe section while a concentrated chlorine solution is added. Chlorine solution is injected using a solution-feed chlorinator or pumped from a concentrated hypochlorite solution (Figure 3-13). This method is acceptable for general application.

Use Tables 3-9a & b to determine the amount of chemical needed to achieve 25 mg/L dosage using the continuous-feed method. When using liquid chlorine from a cylinder, the amounts shown in Table 3-9a are used. The amount for 100 ft and 500 ft of pipeline are listed in the table and the weights are both in pounds and grams. Sodium hypochlorite solution amounts are shown in Table 3-9b. The amount (in both gallons and liters) for 100-ft and 500-ft pipeline lengths and 1% and 5% solution strengths are given.

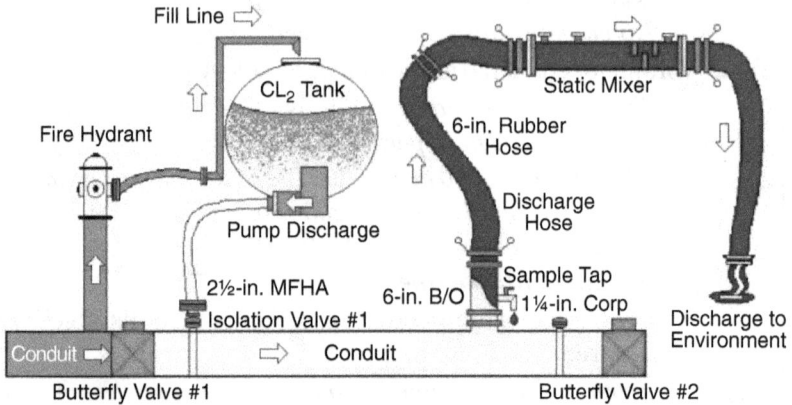

Courtesy Nicole Peschel

Figure 3-13   Pipeline chlorination by continuous-feed method

Chlorination steps and precautions.

1. Flush the new main at a velocity of at least 2.5 ft/sec (Table 3-1) unless site-specific conditions prevent this practice. Large mains (greater than 24 in., or 600 mm) may be broom-swept.

2. Hydrostatic testing must be completed prior to disinfection.

3. Place (optional) calcium hypochlorite granules or tablets in the pipe sections as described in the Tablet Method. This practice provides a strong chlorine concentration during the initial filling of the pipeline and disinfects annular spaces at pipe joints.

4. Supply a constant flow rate from a temporary, backflow-protected source or other approved supply source. Determine the flow using a pitot gauge in the discharge (or other type of meter), measuring the time to fill a precalibrated container, or use the trajectory discharge method (shown in Figure 3-3).

5. Inject the concentrated chlorine solution no more than 10 ft (3 m) from the beginning of the new main. The dosage

should be at least 25 mg/L free chlorine. Tables 3-9a, b list the amount of chlorine required for 100 ft (30.5 m) of pipe of various diameters. Prepare a 1 percent chlorine solution by mixing 1 lb (454 g) of calcium hypochlorite or 3 L of 10 percent strength sodium hypochlorite solution in 8 gal (30.3 L) of water. Use the continuous-feed rate calculators (c3-1 to c3-5) to determine the correct matching feed rate depending on the pipeline fill rate to achieve 25 mg/L chlorine dosage.

6. Measure the chlorine residual regularly to verify the concentration (high-range chlorine test kit or applicable method in current edition of *Standard Method for the Examination of Water and Wastewater* or AWWA Manual M12, *Simplified Procedures for Water Examination*). Take samples for chlorine testing near the point of injection and at the end of the pipeline.

7. Retain the chlorinated water in the main for at least 24 hr. Operate valves and hydrants connected to the treated section to ensure that they are also disinfected.

8. At the end of the 24-hr holding period, the water within the treated main must have a chlorine residual of at least 10 mg/L free chlorine.

9. A solution-feed, vacuum-operated chlorinator (Figure 3-14) and booster pump may be used for chlorine application. Gasoline or electrically powered chemical-feed pumps may be used to inject hypochlorite solutions.

10. Make sure that chlorine solution feed lines are compatible with these chemicals and that the feed system can withstand the feed pressure. The feed system must be free of leaks.

11. Take precautions to control any water discharges (dechlorination may be necessary).

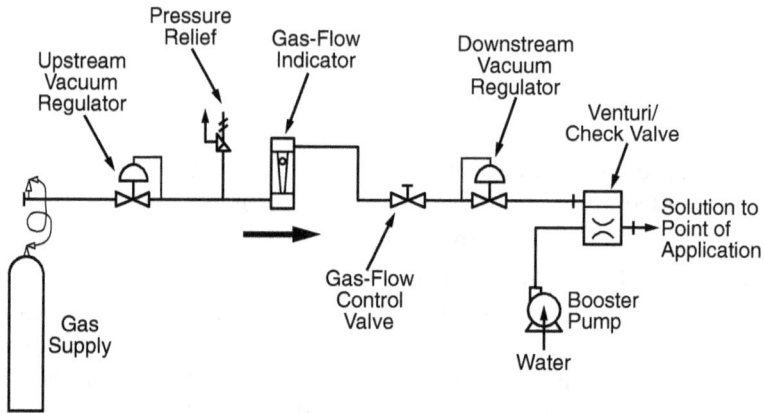

Figure 3-14   Vacuum-operated solution-feed system

Chlorine Amount Calculation

The following tables and calculators should be used to determine the amount of chlorine needed for the continuous-feed method. Chlorine is fed into the water filling the pipeline to achieve a 25 mg/L dosage.

Use the chlorination chemical amount calculation checklist on page 28.

❏ The pipeline diameter and length are used to calculate the volume of the pipeline from Table 3-6.

❏ The number of valves, hydrants, and other appurtenances determine the number of tablets needed for disinfection of these items. Usually one tablet for each is adequate.

❏ This method requires that the entire pipeline is filled with water so the volume of the pipeline is the same as the volume of chlorination water (Table 3-6)

❏ The amount of chlorine chemical needed for the job comes from Tables 3-9a and 3-9b, depending on the length and size of pipe and whether liquid chlorine or sodium hypochlorite are used. The solution or gas chlorinator feed rate is determined using the calculators c3-1, -2, -3, -4, -5. Each calculator equation is for a specific chemical and unit of measure. These chemical feed rates match the fill rate of the pipeline to result in the 25 mg/L dosage required.

Study the examples below to gain a more complete understanding of how to make this calculation.

Table 3-9a   Liquid chlorine required for 25 mg/L dosage

Pipe Diameter		Liquid Chlorine for 100 ft (30.48 m)		Liquid Chlorine for 500 ft (152.4 m)	
in.	*mm**	*lb*	*g*	*lb*	*g*
2	50	0.00	1.54	0.02	7.71
4	100	0.01	6.17	0.07	30.85
6	150	0.03	13.88	0.15	69.42
8	200	0.05	24.68	0.27	123.42
10	250	0.08	38.57	0.42	192.84
12	300	0.12	55.54	0.61	277.69
16	350	0.22	98.74	1.09	493.68
18	400	0.28	124.96	1.38	624.81
20	450	0.34	154.27	1.70	771.37
24	600	0.49	222.16	2.45	1,110.78
30	750	0.76	347.12	3.82	1,735.59
36	900	1.10	499.85	5.50	2,499.24
42	1,100	1.50	680.35	7.49	3,401.75
48	1,200	1.96	888.62	9.79	4,443.10
54	1,400	2.48	1,124.66	12.39	5,623.30
60	1,500	3.06	1,388.47	15.29	6,942.34
64	1,600	3.48	1,579.77	17.40	7,898.84

* The metric equivalents in mm are nominal values, not exact conversions. The chlorine amounts are based on the US pipe diameter figures.

Table 3-9b   Chlorine solution required for 25 mg/L dosage

Pipe Diameter		Amount 1% Solution Needed for*				Amount of 5% Solution Needed for			
		100 ft (30.48 m)		500 ft (152.4 m)		100 ft (30.48 m)		500 ft (152.4 m)	
in.	mm	gal	L	gal	L	gal	L	gal	L
2	50	0.04	0.2	0.20	0.8	0.01	0.03	0.04	0.2
4	100	0.16	0.6	0.80	3.1	0.03	0.12	0.16	0.6
6	150	0.36	1.4	1.79	6.9	0.07	0.28	0.36	1.4
8	200	0.64	2.5	3.19	12.3	0.13	0.49	0.64	2.5
10	250	1.00	3.9	4.98	19.3	0.20	0.77	1.00	3.9
12	300	1.43	5.6	7.17	27.8	0.29	1.1	1.43	5.6
16	350	2.55	9.9	12.75	49.4	0.51	2.0	2.55	9.9
18	400	3.23	12.5	16.13	62.5	0.65	2.5	3.23	12.5
20	450	3.98	15.4	19.92	77.2	0.80	3.1	3.98	15.4
24	600	5.74	22.2	28.68	111.1	1.15	4.4	5.74	22.2
30	750	8.96	34.7	44.82	173.7	1.79	6.9	8.96	34.7
36	900	12.91	50.0	64.54	250.1	2.58	10.0	12.91	50.0
42	1,100	17.57	68.1	87.84	340.4	3.51	13.6	17.57	68.1
48	1,200	22.95	88.9	114.73	444.6	4.59	17.8	22.95	88.9
54	1,400	29.04	112.5	145.20	562.7	5.81	22.5	29.04	112.5
60	1,500	35.85	138.9	179.26	694.7	7.17	27.8	35.85	138.9
64	1,600	40.79	158.1	203.96	790.4	8.16	31.6	40.79	158.1

* Prepare 1% chlorine solution by mixing 1 lb (454 g) calcium hypochlorite or 3 L of 10% sodium hypochlorite solution in 8 gal (30.3 L) of water.

Table Tamer   These tables show the amount of chlorine needed to achieve a 25 mg/L dosage in pipelines of various diameters and in both 100- and 500-ft lengths. The term "liquid chlorine" follows the convention used in ANSI/AWWA Standard B301. Example using Table 3-9b: 500 ft of 24-in. pipeline requires about 5.74 gallons of 5% sodium hypochlorite solution to achieve a 25 mg/L dosage.

Continuous-Feed Method Calculator

To calculate the chlorine feed rate to achieve 25 mg/L inside pipeline (does not account for chlorine demand).

Liquid chlorine feed rate calculations (chlorinator settings).

Liquid chlorine feed rate (lb/day) $=$ 0.3 $\times$ | Water fill rate (gpm) total | (c3-1)

Liquid chlorine feed rate (g/hr) $=$ | Water fill rate (L/sec) total | $\times$ 90   (c3-2)

Chlorine solution feed rate calculations.

Chlorine solution feed rate (mL/min) $=$ 9.46 $\times$ | Water fill rate (gpm) total | $\div$ | Strength of chlorine feed solution (%) | (c3-3)

Chlorine solution feed rate (gal/hr) $=$ 0.15 $\times$ | Water fill rate (gpm) total | $\div$ | Strength of chlorine feed solution (%) | (c3-4)

Chlorine solution feed rate (mL/min) $=$ 150 $\times$ | Water fill rate (L/sec) total | $\div$ | Strength of chlorine feed solution (%) | (c3-5)

Example 1. Using the continuous feed method to disinfect 1,000 ft of 6-in. pipe, liquid chlorine is injected with suction provided by a gas-powered, 20-gpm pump. No other water is being used to fill the pipeline. What is the chlorinator setting to result in 25 mg/L dosage? What is the total amount of chlorine needed for the job?

Use the checklist to ensure all of the information is collected.

❑ The pipeline diameter and length.
   6-in. diameter and 1,000-ft length
❑ Valves, hydrants, tees, and other appurtenances.
   No valves or hydrants in job description
❑ Volume of chlorination water to disinfect pipeline.
   From Table 3-6, 6-in. pipe contains 1.47 gal/ft, so 1,000 ft contains 1,470 gal. At 20 gal/min fill rate, the procedure will take about 74 min.
❑ Amount of chlorine needed to complete the disinfection procedure.
   *From Table 3-9a, 500 ft of 6-in. pipe requires 0.15 lb, so for 1,000 ft, **0.3 lb** is needed.*

☞ Complete the procedure by matching (calculating using c3-1) the liquid chlorine feed rate with the pipeline fill rate (20 gpm in this case).

Liquid chlorine feed rate (lb/day) = 0.3 × Water fill rate (gpm) total

*lb/day = 20 gpm × 0.3 = **6 lb/day feed rate to provide 25 mg/L dosage***

Example 2. Using the continuous feed method to disinfect 500 m of 200-mm pipe, 1 percent sodium hypochlorite solution is injected into water filling the pipeline. The pipeline is filled at 3 L/sec. What is the sodium hypochlorite solution feed rate to result in 25 mg/L dosage? What is the total amount of chlorine needed for the job?

Use the checklist to ensure all of the information is collected.

❑ The pipeline diameter and length.
   200-mm diameter and 500-m length
❑ Valves, hydrants, tees, and other appurtenances.
   No valves or hydrants in job description
❑ Volume of chlorination water to disinfect pipeline.
   *From Table 3-6, 6-in. pipe contains 31.40 L/m, so
   500 m contains 15,700 L. At 3 L/s fill rate, the
   procedure will take about 87 min.*
❑ Amount of chlorine needed to complete the disinfection
   procedure.
   *From Table 3-9b, 153 m of 200-mm pipe requires
   12.3 L of 1 percent sodium hypochlorite solution, so
   for 500 m (12.3 × 500 ÷ 152), 39.5 L is needed.*

☞ Complete the procedure by matching (calculating using c3-5) the chlorine solution feed rate with the pipeline fill rate (3 L/sec in this case).

Chlorine solution feed rate (mL/min) = 150 × Water fill rate (L/sec) total ÷ Strength of chlorine feed solution (%)

*mL/min = 150 × 3 ÷ 1 = 450 mL/min feed rate for
25 mg/L dosage*

Slug Method

The slug method requires that a section of new main is highly chlorinated, then valves are operated and water is slowly removed from the end of the main causing the "slug" of highly chlorinated water to move along the newly installed pipeline (Figure 3-15). As this slug passes tees, crosses, and hydrants, the adjacent valves are operated to ensure the disinfection of all appurtenances and branches.

The slug method generally involves placing calcium hypochlorite granules or tablets in the main during construction; filling

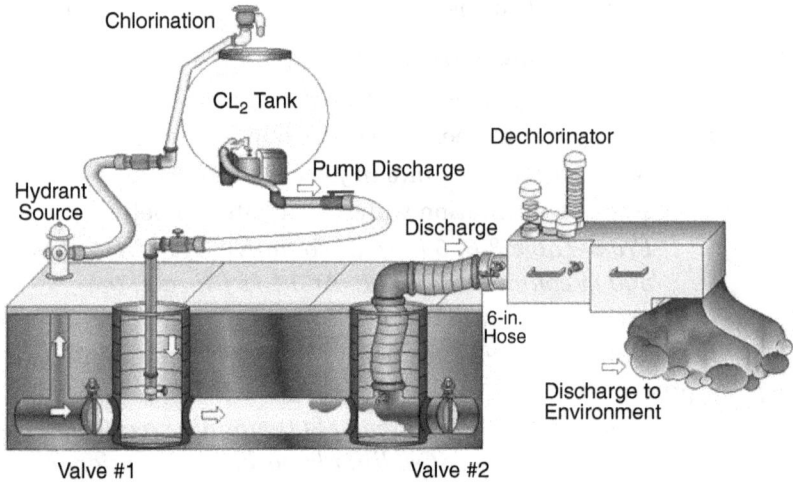

Chlorination

CL$_2$ Tank

Dechlorinator

Pump Discharge

Hydrant Source

Discharge

6-in. Hose

Discharge to Environment

Valve #1   Valve #2

Courtesy Nicole Peschel

Figure 3-15   Slug method pipeline disinfection example

the main to remove air; flushing the main; and then moving a slug of 100 mg/L chlorinated water through the pipeline to ensure at least a 3-hr contact time. The specific steps required for this procedure are listed below.

Chlorination steps and precautions.

1. Flush the new main at a velocity of at least 2.5 ft/sec (Table 3-1) unless site-specific conditions prevent this practice. Large mains (greater than 24 in., or 600 mm) may be broom-swept.

2. Hydrostatic testing must be completed prior to disinfection.

3. During construction, place (optional) calcium hypochlorite granules or tablets in the pipe sections as described in the Tablet Method using Tables 3-7a, 3-7b, 3-8a, and 3-8b to determine the amount of granules and tablets needed. This practice provides a strong chlorine concentration during initial filling of the pipeline and disinfects annular spaces at pipe joints. Do not use this procedure on solvent-welded plastic or on screwed-joint steel pipe.

4. Fill the pipeline containing the calcium hypochlorite granules or tablets slowly (not more than 1 ft/sec). Remove air pockets. Flush this chlorinated water from the pipeline. Dechlorinate this water is necessary.

5. Supply a constant flow rate, to fill the pipeline and when forming and moving the slug, from a temporary, backflow-protected source or other approved supply source. Determine the flow using a pitot gauge in the discharge (other type of meter), measuring the time to fill a precalibrated container, or use the trajectory discharge method (shown in Figure 3-3).

6. Inject the concentrated chlorine solution no more than 10 ft (3 m) from the beginning of the new main. The dosage should be at least 100 mg/L free chlorine. Tables 3-10, 3-11, and 3-12 list the liquid chlorine, sodium hypochlorite, and calcium hypochlorite application rates to achieve this dosage. Prepare a 1 percent chlorine solution by mixing 1 lb (454 g) of calcium hypochlorite or 3 L of 10 percent strength sodium hypochlorite solution in 8 gal (30.3 L) of water.

7. Measure the chlorine residual (near the injection point) regularly to verify the concentration (high-range chlorine test kit or applicable method in current edition of *Standard Method for the Examination of Water and Wastewater* or AWWA Manual M12).

8. Apply chlorine continuously to develop a solid column, or slug, of chlorinated water that will expose all interior surfaces to a concentration of approximately 100 mg/L for at least 3 hr. Measure the chlorine residual at the end of the slug to verify the concentration.

9. Measure the free chlorine residual as the slug moves through the new pipeline (Table 3-12). If the residual drops below 50 mg/L at any point, stop the procedure and add chlorine to the slug area to restore the free chlorine concentration to at least 100 mg/L. Choose sample locations that will verify the minimum contact time of 3 hr.

10. Operate valves and hydrants as the slug passes them to disinfect appurtenances and pipe branches.

11. High calcium hypochlorite concentrations (greater than 100 mg/L) may raise the pH of the water to unacceptable levels. Measure the pH of the chlorinated water to ensure that optimum pH is maintained. Remember that if the pH is greater than 10, virtually all of the chlorine residual is present as hypochlorite ion, which is a less effective disinfectant than hypochlorous acid.

Chlorine amount calculation. The following tables and calculators should be used to determine the amount of chlorine and the application circumstances to correctly perform the slug method of pipeline chlorination. A section of the pipeline is filled with 100 mg/L chlorine water (the slug), and this portion is moved along the length of the pipe so that there is a 3-hr contact at every point. The size of the slug and the rate of movement are critical to the success of this method.

The chlorinator feed settings needed to produce a 100 mg/L dosage at various flow rates are given in Table 3-10. These settings are not exact but should yield a dosage suitable for field applications. The flow rate given in the left column is the "fill rate" of the pipeline. So, if the pipeline is being filled at 20 gpm, the chlorinator should be set at 24 lb/day to produce a 100 mg/L slug. Using SI units, if the pipeline is being filled at 2 L/sec, the chlorinator should be set at 720 grams/hr.

Table 3-11 lists the feed rate needed to prepare a 100 mg/L dosage slug using sodium hypochlorite solutions. The feed rate is matched with the pipeline feed (or fill) rate. From the table, if the pipeline is being filled at 25 gpm, the feed rate of a 1% sodium hypochlorite solution must be about 15 gallons/hr.

The amount of calcium hypochlorite needed to produce a 100 mg/L dosage for each 500 ft of pipeline is given in Table 3-12. The amount is listed in ounces (pounds), grams (kilograms), and the number of 5-gram tablets. Tablets are generally not used for this method since a large number is usually needed. It is common to use calcium hypochlorite to prepare 1% or 5% chlorine solutions to produce the slug as shown in Table 3-11. A 1% solution is prepared by mixing 1 pound of calcium hypochlorite (65% strength) in 8 gallons of water.

Table 3-10   Approximate feed rate settings for 100 mg/L chlorine dosage at various flow rates (liquid chlorine feed)

US Units		SI Units	
Flow Rate *gpm*	Feed Rate Setting *lb/day*	Flow Rate *L/sec*	Feed Rate Setting *g/hr*
10	12	0.5	180
20	24	1	360
30	36	1.5	540
40	48	2	720
50	60	2.5	900
60	72	3	1,080
70	84	3.5	1,260
80	96	4	1,440
90	108	4.5	1,620
100	120	5	1,800

Table Tamer   This table shows the approximate chlorinator feed rate settings for various fill rates to achieve 100 mg/L dosage. To get the setting for other dosages, just multiply the setting for 100 mg/L by the ratio: other dosage/100 mg/L. Example: The pipeline fill rate is 30 gpm but the desired dosage is 25 mg/L. For a 30 gpm fill rate the chlorinator feed setting from the table is 36 lb/day. Multiply this setting times the ratio 25/100 = 9 lb/day.

Use the chlorination chemical amount calculation checklist on page 28 to ensure that you have all of the items needed to make the calculation.

❑ The pipeline diameter and length are used to calculate the volume of the pipeline from Table 3-6.

❑ The number of valves, hydrants, and other appurtenances determine the number of tablets needed for disinfection of these items. Usually one tablet for each is adequate.

Table 3-11   Approximate hypochlorite solution feed rates at various flow rates for 100 mg/L dosage

US Units			SI Units		
Water	Solution Feed Rate		Water	Solution Feed Rate	
Fill Rate gpm	1% Solution gph	5% Solution gph	Fill Rate L/sec	1% Solution* mL/min	5% Solution* mL/min
1	0.6	0.12	0.1	60	12
2	1.2	0.24	0.2	120	24
3	1.8	0.36	0.3	180	36
4	2.4	0.48	0.4	240	48
5	3	0.6	0.5	300	60
6	3.6	0.72	0.6	360	72
7	4.2	0.84	0.7	420	84
8	4.8	0.96	0.8	480	96
9	5.4	1.08	0.9	540	108
10	6	1.2	1	600	120
15	9	1.8	1.1	660	132
20	12	2.4	1.2	720	144
25	15	3	1.3	780	156
30	18	3.6	1.4	840	168
35	21	4.2	1.5	900	180
40	24	4.8	1.6	960	192
45	27	5.4	1.7	1,020	204
50	30	6	1.8	1,080	216
55	33	6.6	1.9	1,140	228
60	36	7.2	2	1,200	240
65	39	7.8	2.5	1,500	300
70	42	8.4	3	1,800	360
75	45	9	3.5	2,100	420
80	48	9.6	4	2,400	480
85	51	10.2	4.5	2,700	540

Table continued next page.

Table 3-11   Approximate hypochlorite solution feed rates at various flow rates for 100 mg/L dosage (continued)

US Units			SI Units		
Water Fill Rate gpm	Solution Feed Rate 1% Solution gph	5% Solution gph	Water Fill Rate L/sec	Solution Feed Rate 1% Solution* mL/min	5% Solution* mL/min
90	54	10.8	5	3,000	600
95	57	11.4	5.5	3,300	660
100	60	12	6	3,600	720
110	66	13.2	6.5	3,900	780
120	72	14.4	7	4,200	840
130	78	15.6	7.5	4,500	900
140	84	16.8	8	4,800	960
150	90	18	8.5	5,100	1,020
160	96	19.2	9	5,400	1,080
170	102	20.4	9.5	5,700	1,140
180	108	21.6	10	6,000	1,200
190	114	22.8	10.5	6,300	1,260
200	120	24	11	6,600	1,320

* Chlorine solution is prepared by mixing 1 lb calcium hypochlorite (65%) with 8 gal of water or diluting 10% sodium hypochlorite 1 gal with 9 gal of water. Five percent chlorine solution is prepared by mixing 5 lb calcium hypochlorite (65%) with 8 gal of water or diluting 10% sodium hypochlorite 1 gal with 1 gal of water.

Table Tamer   Table shows the approximate feed rate for 1% and 5% hypochlorite solutions to match various fill rates using the slug method. Example: Feed a 1% hypochlorite solution at 12 gph to match a 20 gpm fill rate and to achieve a 100 mg/L chlorine dosage. See shaded cell on previous page.

Table 3-12   Total chlorination time needed based on slug size

Slug Size	10%	15%	20%	25%	30%	35%	40%	45%	50%
Chlorination time, *hr*	30	22.5	15	12.5	10	8.75	7.5	6.75	6

❑ This method uses a slug of 100 mg/L chlorinated water as a slug that moves along the length of the pipeline. The slug is a percentage of the total length of the pipeline. The slug volume is calculated from Table 3-6 and calculator c3-6.

❑ The amount of chlorine chemical needed for the job is calculated using c3-8, 9, or 10, depending on the length and size of pipe and the chemical used. In the situation where the slug is produced by continuously feeding the chlorination chemical, the withdrawal rate is calculated from c3-7 to achieve a 3-hr contact time at every point along the pipeline. Tables 3-10 and 3-11 show the feed rate needed to match with the fill rate (withdrawal rate) of the pipeline.

Study the examples below to gain a more complete understanding of how to make this calculation.

Volume of slug calculator.   Choose the size of the slug (percent of total volume) based on the time to complete the job (Table 3-12). For example, if only 10 hr are available to complete the chlorination, a 30 percent slug would be the minimum size necessary.

For the slug method, calculate the withdrawal rate using the following calculator to ensure 3-hr contact time at all points along the pipeline. Choose the matching chlorination chemical solution feed rate from Table 3-10 or 3-11.

Table 3-12 shows the total time needed to complete the disinfection for various sizes of slugs. If the slug is 30% of the length of the pipeline to be disinfected, the procedure will take 10 hours (including creating the slug and then moving the slug so that there is a 3-hr contact time). This table is helpful to select the size of the slug based on the time needed to complete the job.

Volume of Slug Calculator.

$$
\begin{array}{c}
\text{Volume} \\
\text{of slug}
\end{array}
=
\boxed{\begin{array}{c}\text{Total length}\\\text{of pipe to be}\\\text{disinfected}\end{array}}
\times
\boxed{\begin{array}{c}\text{Volume/ft or}\\\text{Volume/m}\\\text{from}\\\text{Table 3-5}\end{array}}
\times
\boxed{\begin{array}{c}\text{Slug}\\\text{fraction =}\\\text{decimal}\\\text{fraction of}\\\text{total volume}\end{array}}
\quad (\text{c3-6})
$$

Withdrawal rate calculator.

$$
\begin{array}{c}
\text{Withdrawal rate or fill}\\
\text{rate for 3 hr contact}\\
\text{(gpm or L/min)}
\end{array}
=
\boxed{\begin{array}{c}\text{Volume of slug}\\\text{(from c3-6)}\end{array}}
\div \quad 180 \text{ min}
\quad (\text{c3-7})
$$

Ops Tip   180 min in 3 hr is used here since withdrawal rate is in minutes.

Chlorination chemical amount calculator.

$$
\begin{array}{c}
\text{Volume of 1\%}\\
\text{hypochlorite solution}\\
\text{needed for slug}
\end{array}
=
\boxed{\begin{array}{c}\text{Volume of slug}\\\text{(from c3-6)}\end{array}}
\times \quad 0.01
\quad (\text{c3-8})
$$

$$
\begin{array}{c}
\text{Amount of liquid}\\
\text{chlorine needed for}\\
\text{slug, lb}
\end{array}
=
\boxed{\begin{array}{c}\text{Volume of slug}\\\text{(from c3-6), gal}\end{array}}
\times \quad 0.00083
\quad (\text{c3-9})
$$

$$
\begin{array}{c}
\text{Amount of liquid}\\
\text{chlorine needed for}\\
\text{slug, kg}
\end{array}
=
\boxed{\begin{array}{c}\text{Volume of slug}\\\text{(from c3-6), L}\end{array}}
\times \quad 0.0001
\quad (\text{c3-9})
$$

Example 1.   Using the slug method, chlorinate a 1,000-ft length of 6-in. diameter pipeline. There are four fire hydrants and two valves located on the new section of pipeline. The chlorination must be completed in less than 15 hr. A 1 percent sodium hypochlorite solution is used for this procedure. How much is needed and describe the process.

Use the checklist to ensure all of the information is collected.

❑ The pipeline diameter and length.
   6-in. diameter and 1,000-ft length
❑ Valves, hydrants, tees, and other appurtenances.
   Four hydrants and two valves. Chlorinate each with one calcium hypochlorite tablet.
❑ Volume of chlorination water to disinfect pipeline.
   Select a 25% slug from Table 3-12 because less than 15 hr is available. A 25% slug will take 12.5 hr. The volume of the slug is calculated from c3-6.
   *1,000 × 1.47 × 0.25 = **367.5 gallons***
❑ Amount of chlorine needed to complete the disinfection procedure.
   1% sodium hypochlorite feed rate to match withdrawal rate of 2 gpm is 1.2 gph (from Table 3-11). The withdrawal rate and the fill rate are the same. It takes 3 hr to create the 10% slug, so 1.2 gph × 3 hr = 3.6 gal of 1% sodium hypochlorite is needed for the job.
☞ Calculate the withdrawal rate using c3-7.
   *367.5 ÷ 180 = **2.0 gpm for 3-hr contact time along the pipeline.***

Example 2.   Use the slug method to chlorinate 2,000 m of 600-mm diameter pipe. Use 5 percent hypochlorite solution (prepared from calcium hypochlorite) for the procedure. Up to 16 hr are available for the job. How much solution is needed? Describe the procedure.

Use the checklist to ensure all of the information is collected.

- ❑ The pipeline diameter and length.

  600-mm diameter and 2,000-m length
- ❑ Valves, hydrants, tees, and other appurtenances.

  No valves or hydrants are listed in this job.
- ❑ Volume of chlorination water to disinfect pipeline.

  Select a 20% slug from Table 3-12 because 16 hr are available and a 20% slug will take 15 hr to complete. The volume of the slug is calculated from c3-6.

  *2,000 × 282.6 × 0.2 = **113,040 L***
- ❑ Amount of chlorine needed to complete the disinfection procedure.

  5% sodium hypochlorite feed rate to match withdrawal rate of 628 L/min or 10.5 L/s is 1,260 mL/min (from Table 3-11). The withdrawal rate and the fill rate are the same. It takes 3 hr (180 min) to create the 20% slug, so 1,260 × 180 = 226,800 mL or 226.8 L of 5% sodium hypochlorite is needed for the job.
- ☞ Calculate the withdrawal rate using c3-7.

  *113,040 ÷ 180 = **628 L/min for 3-hr contact time along the pipeline.***

References

ANSI/AWWA C651. *AWWA Standard for Disinfecting Water Mains.* American Water Works Association: Denver, Colo.

Standard Methods for the Examination of Water and Wastewater. (current edition). APHA, AWWA, WEF: Washington, D.C.

Tikkanen, M., and J. Schroeter. 2001. *Guidance Manual for Disposal of Chlorinated Water,* Awwa Research Foundation: Denver, Colo.

Water Transmission and Distribution, 3rd ed. 2004. American Water Works Association: Denver, Colo.

Zhang, Jian. *Guidelines to Minimize Downtime During Pipe Lining Operations,* Awwa Research Foundation preliminary draft. University of Louisville, AWWA: Denver, Colo.

Chapter 4

Disinfection for Pipeline Repairs

Broken pipelines, severe service line leaks, and other damaged distribution system components require emergency repairs. An action plan is needed to efficiently handle these situations. The plan should include coordination with police, fire, and street or highway department personnel, and describe personnel training requirements, discuss the need for accurate records, and ensure the availability of critical repair parts. Careful planning can lead to the quick restoration of water service.

When making emergency repairs, the following steps should be used to reduce the chance of microbial contamination and to safely return the pipeline to service.

1. Ensure that repair parts are available and in good sanitary condition before they arrive at the site.
2. Employ sanitary repair practices and check the trenches for potential contamination.
3. Use accepted chlorination procedures.
4. Take samples for bacteriological testing.
5. Document the event and the procedures used to repair and return the pipeline to service.

Each of these good practices are discussed in more detail in the following sections.

Condition and Storage of Repair Parts

A carefully selected inventory of spare (repair) parts should be stored so that they can be delivered quickly to the site of an emergency main break. Either a parts supplier or the utility property are suitable sites for locating a parts inventory, as long as they are accessible when they are needed.

All pipes, fittings, valves, clamps, and other appurtenances should be inspected on delivery. Security should be provided to protect these materials while they are in storage. Pipe and other parts should be stored in a manner to avoid casual water and dirt from entering. Any visible foreign matter should be removed from the interior of stored repair parts. Watertight end caps or other protective coverings should be used and maintained to prevent possible contamination.

Sanitary Repair Practices

Care should be taken to prevent water, dirt, and other material from entering the damaged pipeline. Surface water should be diverted from the construction site using barriers if necessary. The excavation should be dewatered so that water is below the pipe invert (Figure 4-1). Before beginning a repair, the interior of the pipe should be cleaned where it has contacted the soil or backfill material. Soil beyond reach can be removed by low velocity flushing in both directions. If possible, the flow or positive pressure should be maintained in the leaking pipeline to prevent backflow of contaminated water. This is particularly important until the leak is exposed and secured.

> **Ops Tip**   Take special care to keep the pipe clean. This will make disinfection more effective.

Repair crews need to be aware that their actions could contaminate the water supply during a pipeline repair. Crews must ensure the cleanliness of all cables, pipes, and hoses drawn through the inside of the pipelines or appurtenances. Disinfection (using chlorine dip or swab solutions) of hand tools (such as saws) used in the repair is a prudent step. Equipment operators should also use caution to avoid introducing soil or other material into the repaired pipeline.

Chlorination Procedures

When an existing main is opened because of a break or leak, the excavation is usually wet and may be contaminated. ANSI/AWWA Standard C651, *Disinfecting Water Mains*, requires the application of liberal amounts of hypochlorite to the open trench

Courtesy Denver Water
Figure 4-1   Excavation dewatering for pipeline repair

to reduce potential pollution. Calcium hypochlorite tablets are used for this application because they dissolve slowly, continuously releasing chlorine as the repair is completed.

Chlorine solution (1 percent hypochlorite prepared according to Table 3-11) should be applied to the interior of pipe and fittings used for pipeline repair. The solution can be swabbed with an applicator or sprayed (Figure 4-2) on the interior surfaces of the pipeline and repair parts.

The section of repaired main should be disinfected whenever possible. The slug method (Chapter 3) is usually used because it takes less time to accomplish. The repaired section is isolated, all service connections shut off, and then the main is flushed before chlorinating. The chlorine dose may be as much as 300 mg/L (Table 4-1) and the contact time as short as 15 min. Calcium hypochlorite granules or tablets require some time to dissolve. Therefore, this chemical may not be the best choice if the main

Courtesy Nicole Peschel
Figure 4-2   Sprayers for repair part disinfection

must be returned to service quickly. The repaired main is flushed (dechlorinate if necessary) until the chlorine residual is no higher than the distribution system feed water (distribution systems that use chloramines require testing for free and combined chlorine).

The repaired main should be thoroughly flushed regardless of whether it is chlorinated or not. Flushing is the best way to remove any foreign material that may have entered the main during the repair. The water should be flushed until it is clear and the chlorine residual is equivalent to the distribution system feed water.

> **Ops Tip**   Flushing the repaired main before returning it to service is
> ☼   critical.

Tapping sleeves are used to install a tap in the main without the need to shut down the main. When using these devices, the exterior of the main should be cleaned in the tap location. Also, calcium hypochlorite should be applied to the interior of the sleeve. After the tap is made, it is impossible to disinfect the

Table 4-1   Calcium hypochlorite granules or tablets needed for 100 mg/L and 300 mg/L dosage for 100 ft (30.48 m) length

Pipe Diameter		Calcium Hypochlorite Amount					
		100 mg/L Dosage			300 mg/L Dosage		
in.	mm	oz	g	Tablets	oz	g	Tablets
2	50	0.3	10	2	1	29	6
4	100	1.3	38	8	4	114	23
6	150	3.0	86	17	9	257	51
8	200	5.4	152	30	16	456	91
10	250	8.4	238	48	25	713	143
12	300	12	342	68	36	1,027	205
		Change in Units			Change in Units		
		lb	kg		lb	kg	
14	350	1	0.5	93	3	1	280
16	400	1	0.6	122	4	2	365
18	450	2	0.8	154	5	2	462
20	500	2	1.0	190	6	3	571
24	600	3	1.4	274	9	4	822
30	800	5	2.1	428	14	6	1,284
36	900	7	3.1	616	20	9	1,849
42	1,000	9	4.2	839	28	13	2,516
48	1,200	12	5.5	1,095	36	16	3,286
54	1,400	15	6.9	1,386	46	21	4,159
60	1,500	19	8.6	1,712	57	26	5,135
64	1,600	21	9.7	1,947	64	29	5,842

Table Tamer   The units used in this table are changed from ounces to pounds and grams to kilograms in the middle.

annulus without a pipeline shutdown. The space between the sleeve and the pipe is only about ½ in., so a very small amount of calcium hypochlorite applied to the interior of the sleeve is all that is necessary to result in a high chlorine dosage.

Bacteriological Testing

After the main is placed back into service (some regulatory agencies may require satisfactory bacteriological test results *before* the main is placed back into service), samples should be taken from the repaired main as described in Chapter 3 and tested for total coliform bacteria. In most cases, samples should be taken from both upstream and downstream of the repair. If a positive total coliform test is obtained, corrective action should be taken. Actions may include additional flushing or chlorination of the repaired section. ANSI/AWWA Standard C651 requires continued sampling daily until two consecutive negative results are obtained.

Documentation

Accurate records should be kept for future reference. The date, location, type of break, materials used for the repair, personnel involved in the repair, procedures used, and test results should be recorded. These records are useful if problems develop or to determine a pattern of pipeline breaks in a specific area.

References

ANSI/AWWA C651. *AWWA Standard for Disinfecting Water Mains.* American Water Works Association: Denver, Colo.

Good Practices for Preventing Microbiological Contamination of Water Mains. 2001. American Water Works Association: Denver, Colo.

Standard Methods for the Examination of Water and Wastewater. (current edition). APHA, AWWA, WEF: Washington, D.C.

Chapter 5

Disinfection of Storage Facilities

Disinfection of newly constructed potable water storage facilities (e.g., tanks, standpipes, underground basins, covered reservoirs) is required. Additionally, all storage facilities taken out of service for inspection or maintenance must be disinfected before they are returned to service. The procedures discussed in this chapter follow the requirements of ANSI/AWWA C652, *Standard for Disinfection of Water-Storage Facilities*. The standard should be consulted for specific language and other requirements. Applicable regulatory agencies should be contacted to determine if there are local requirements that apply to the disinfection of potable water storage facilities.

The following steps should be used to properly disinfect potable water storage facilities.

1. Use sanitary construction methods.
2. Employ approved chlorination (chemical disinfection) procedures.
3. Remove and properly dispose of highly chlorinated water (full storage facility chlorination method).
4. Perform bacteriological testing.
5. Document the release to potable service.

Each of these good practices are discussed in more detail in the following sections.

Sanitary Construction or Maintenance Practices

All materials not part of the structure or operating equipment should be removed. This may include mold on walls or ceilings, and algae attached to surfaces. The interior walls, floors, and other items that are part of the structure should be cleaned with high-pressure water, sweeping, scrubbing, or other effective methods. Material removed by cleaning procedures must be completely removed. All penetration screens should be checked to ensure that openings are protected. The ceiling of the storage facility should be checked for cracks (look for daylight). Finally, any materials that are placed in the facility following the cleaning must be clean and sanitary.

Chlorination Safety

All safety precautions as indicated by the Chlorine Institute should be followed when performing chlorination procedures. Working on or in a storage facility can be hazardous. There may be regulatory requirements that apply to ensure that these procedures are conducted in a safe manner. Appropriate regulatory agencies and requirements should be consulted before beginning any chlorination procedure. Some precautions include:

- Personal protective equipment, such as goggles, gloves, and breathing apparatus.
- Special fans or other ventilation equipment for use within storage facilities.
- Adequate lighting.
- Electrical equipment that is specifically designed for safe use in a wet environment.
- Protective clothing.
- Safety cages or cables to prevent accidents on ladders or entry steps.
- Confined space entry regulatory requirements.

Removal of Highly Chlorinated Water

Discharge regulations (applicable locally) may dictate the methods of disposal for highly chlorinated water. This is particularly applicable to the full storage facility chlorination method. When using this method, the entire volume of the storage facility contains water with a chlorine residual greater than 10 mg/L. Environmental protection regulations may require special provisions or permits prior to disposal of this water. Dechlorination may be required (methods are described in Chapter 7). The proper authorities should be contacted prior to disposal to determine all applicable regulations and requirements. The applicable waste-water treatment facility should be notified when discharging to a sanitary sewer.

Bacteriological Testing

Water from the storage facility must be sampled following completion of the chlorination method and should be tested for total coliform bacteria according to the latest edition of *Standard Methods for the Examination of Water and Wastewater* (chlorine residual, free and combined, is tested to verify conformance with the chlorination requirements and to restore the water to potable quality). If the test is negative for total coliform, the facility can be released and placed into service. If the test is positive for total coliform, repeated samples should be taken until two consecutive samples (taken 24 hr apart) are negative. The facility may be rechlorinated according to the previously stated procedures and then tested again for total coliform as previously described.

Samples should be taken from a sample tap on the outlet piping or from a sample tap that is directly connected to the facility. The water drawn from the tap must represent the water within the facility. Additional samples may be advisable to ensure uniform disinfection results. Storage facility hatches are suitable for this purpose. It is recommended to sample the facility fill-water to make sure it does not contain total coliform bacteria.

Additional Test for Odor

The water in the facility should be tested for odor. Offensive odors may indicate the presence of trace substances that may cause customer complaints. If an offensive odor is detected, steps should be taken to eliminate it before the facility is placed in ser-vice. This may include further cleaning or allowing additional time for odors to dissipate before returning the tank to service.

Release to Potable Service

The disinfection procedure should be documented and required records submitted to applicable regulatory agencies. The facility should be connected to the distribution system and the system operators notified that the facility if now active. If disinfection is conducted by contractors, full documentation of all aspects of the disinfection process should be required and test results certified.

Chlorination Methods

There are three methods for the chlorination of water storage facilities:

1. The full storage facility chlorination method.
2. The surface application method.
3. The chlorinate and fill method.

A summary of the major features of each method is listed in Table 5-1. Generally, the full storage facility method is used for smaller size tanks. Both the surface application and chlorinate and fill methods are commonly used for larger facilities.

Table 5-1   Storage facility disinfection methods summary

Method	Chlorine Dosage	Holding Time	Chlorine Residual	Notes
Full Storage Facility	10 mg/L	6 hr if continuously fed during filling or 24 hr if two-step process is used	10 mg/L	Repeat procedure if residual is less than 10 mg/L, reduce residual prior to using
Surface Application	200 mg/L, add to drains to achieve 10 mg/L when filled	30 min	10 mg/L for drains	Remove water from drains and fill with potable water
Chlorinate and Fill	50 mg/L in about 5% of the volume	6 hr, then fill with potable water and hold for another 24 hr	2 mg/L in full facility after holding period	Remove water from drains before filling

Full Storage Facility Chlorination Method

This method generally involves chlorinating the entire volume of the storage facility so that the water within the facility has a residual chlorine concentration of at least 10 mg/L at the end of the required retention period. Liquid chlorine, sodium hypochlorite solution, or calcium hypochlorite tablets (or granules) are suitable for this procedure.

The chlorination process may involve a continuous fill of the facility with chlorinated water with a concentration of at least 10 mg/L. Alternatively, the chlorination can be performed in two steps: (1) adding the entire amount of chlorine to disinfect the facility to a small amount of water covering the floor, and (2) filling (after a holding period) the facility with potable water to achieve 10 mg/L throughout.

Liquid Chlorine

Liquid chlorine should be fed into the water filling the facility so that the concentration is constant. Portable equipment includes a liquid chlorine cylinder, a gas-flow chlorinator, a chlorine ejector, safety equipment, and appropriate connections to inject the chlorine solution into the fill stream. Trained operators are needed for this operation. Safety regulations may require personnel certification for emergency response and first-aid training.

Sodium Hypochlorite

When using sodium hypochlorite solution, it should be added to the feed water with a chemical feed pump or by pouring it into the fill-stream. In each case, the solution is added to ensure adequate mixing and to result in a constant chlorine concentration. Convenient locations to pour sodium hypochlorite are the cleanout or inspection manhole in the lower level of the facility, the riser pipe of an elevated tank, or a roof manhole (Figure 5-1). The hypochlorite solution should be poured in when the water is 1–3 ft deep, if possible.

Calcium Hypochlorite

Granules or tablets broken into small pieces (¼ in. or smaller) can be added to the facility as previously described for the sodium

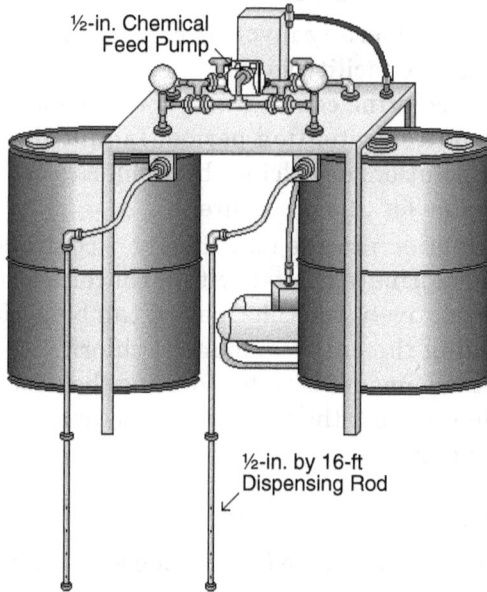

Courtesy Nicole Peschel
Figure 5-1   Sodium hypochlorite feed equipment example

hypochlorite solution. The granules or tablets should be placed on a dry surface unless safety precautions are taken to ensure adequate ventilation or provide personal protective equipment.

Chlorination Steps and Precautions

1. Retain the chlorinated water in the facility for at least 6 hr when continuous feed is used from the gas-chlorinator or chemical feed pump. If a high chlorine concentration is added by pouring sodium hypochlorite or by placing calcium hypochlorite granules or tablets and filling the storage facility, the water should be retained for at least 24 hr. If a two-step process is used, the tablets or granules are allowed to dissolve in a small amount of water (a few feet) before the facility is filled to capacity. The chemical may take several hours to dissolve.

2. Measure the chlorine residual of the water following the required retention period. The free chlorine residual must be at least 10 mg/L. If the chlorine residual is below 10 mg/L, add more chlorine and retain for another period as described in step 1. Continue this procedure until a free chlorine residual of at least 10 mg/L is obtained following the required retention period.

3. Reduce the chlorine residual of the water within the storage facility by draining (dechlorinate if necessary) and refilling with potable water. Another method of achieving this result is to hold the water until the chlorine residual is reduced and blend the remaining water with potable water.

Chlorination Amount Calculation

The following calculators (c5-1, 5-2, 5-3, 5-4, 5-5) and tables (5-2, 5-3) should be used to calculate the amount of chlorination chemical needed to disinfect tanks using the full storage facility chlorination method. These calculations generally involve calculating the volume of the storage facility (using the *tank volume calculators or Table 5-2 for horizontal cylindrical tanks*) and determining the amount of chlorination chemical needed *(using the chlorination chemical calculator)* to achieve a 10 mg/L concentration in that volume.

The volume of most tanks is calculated using the volume calculators below for rectangular, square, vertical cylindrical, and horizontal cylindrical tanks. Other shapes will require information from the manufacturer or design engineer. Table 5-2 shows the volume of horizontal cylindrical tanks with some common dimensions.

The amount of chlorination chemical needed for a unit volume (either 100,000 gallons or 100 cubic meters) to produce 10 mg/L dosage is listed in Table 5-3. The values in this table are then used in the chlorination chemical calculator (c5-5). All three of the most common chlorination chemicals are included in the table with both US and SI units of measure.

Tank Volume Calculators

Rectangular or Square Tanks

$$\boxed{\begin{array}{c}\text{Length,}\\\text{in ft}\end{array}} \times \boxed{\begin{array}{c}\text{Width,}\\\text{in ft}\end{array}} \times \boxed{\begin{array}{c}\text{Maximum}\\\text{depth of}\\\text{water, in ft}\end{array}} \times 7.48 \text{ gal/ft}^3 = \boxed{\begin{array}{c}\text{Volume,}\\\text{in gal}\end{array}} \quad \text{(c5-1)}$$

$$\boxed{\begin{array}{c}\text{Length,}\\\text{in m}\end{array}} \times \boxed{\begin{array}{c}\text{Width,}\\\text{in m}\end{array}} \times \boxed{\begin{array}{c}\text{Maximum}\\\text{depth of}\\\text{water, in m}\end{array}} = \boxed{\begin{array}{c}\text{Volume,}\\\text{in m}^3\end{array}} \quad \text{(c5-2)}$$

Vertical (upright) Cylindrical Tanks

$$5.87 \times \boxed{\begin{array}{c}\text{Diameter,}\\\text{in ft}\end{array}} \times \boxed{\begin{array}{c}\text{Diameter,}\\\text{in ft}\end{array}} \times \boxed{\begin{array}{c}\text{Maximum}\\\text{depth of}\\\text{water, in ft}\end{array}} = \boxed{\begin{array}{c}\text{Volume,}\\\text{in gal}\end{array}} \quad \text{(c5-3)}$$

$$0.785 \times \boxed{\begin{array}{c}\text{Diameter,}\\\text{in m}\end{array}} \times \boxed{\begin{array}{c}\text{Diameter,}\\\text{in m}\end{array}} \times \boxed{\begin{array}{c}\text{Maximum}\\\text{depth of}\\\text{water, in m}\end{array}} = \boxed{\begin{array}{c}\text{Volume,}\\\text{in m}^3\end{array}} \quad \text{(c5-4)}$$

Horizontal (On Side) Cylindrical Tanks

Calculating the volume of partially full horizontal cylindrical tanks is more complex than the same calculation for vertical cylindrical tanks. Table 5-2 shows the volume of common sizes of these tanks. Other tank dimension volumes can be estimated from these values. This type of tank is used primarily for smaller volumes.

Table 5-2   Approximate volume of partially full horizontal cylindrical tank (with hemispheres on both ends)

US Units				SI Units			
Diameter ft	Depth to overflow ft	Length ft	gal	Diameter m	Depth to overflow m	Length m	m³
2	1.5	10	215	0.5	0.4	3	0.56
2	1.5	15	310	0.5	0.4	5	0.9
2	1.5	20	405	0.5	0.4	7	1.24
3	2.5	10	568	1	0.9	3	2.74
3	2.5	15	804	1	0.9	5	4.23
3	2.5	20	1,040	1	0.9	7	5.7
4	3.5	10	1,112	1.5	1.4	3	6.89
4	3.5	15	1,548	1.5	1.4	5	10.3
4	3.5	20	1,982	1.5	1.4	7	13.8
4	3.5	25	2,421	2	1.9	3	13.4
5	4.5	10	1,868	2	1.9	5	19.6
5	4.5	15	2,564	2	1.9	7	25.7
5	4.5	20	3,261	2	1.9	10	35
5	4.5	25	3,957	–	–	–	–
5	4.5	30	4,653	–	–	–	–

Table Tamer   This table lists the volume of horizontal cylindrical tanks with common dimensions. Example: A horizontal cylindrical tank that is 3 ft in diameter and 15 ft long with a depth to overflow of 2.5 ft contains about 804 gallons.

Table 5-3   Approximate amount of chlorination chemicals needed for 10 mg/L concentration

Volume Treated	Liquid Chlorine		5% Sodium Hypochlorite		65% Calcium Hypochlorite	
	lb	*kg*	*gal*	*L*	*lb*	*kg*
100,000 gal	8.3	3.8	19.4	73.4	12.8	5.8
100 m³	2.2	1.0	5.1	19.4	3.4	1.5

Chlorination Chemical Calculator

The approximate amount of chlorination chemical needed for 10 mg/L chlorine concentration for a given size tank is calculated as follows (use either US or SI units):

$$
\underset{1}{\boxed{\begin{array}{c}\text{Volume}\\\text{(from tank}\\\text{shape}\\\text{equations)}\end{array}}} \times \underset{2}{\boxed{\begin{array}{c}\text{Amount}\\\text{of}\\\text{chemical}\\\text{(from}\\\text{Table 5-3)}\end{array}}} \times \underset{3}{\boxed{\begin{array}{c}\text{Chemical}\\\text{strength}\\\text{factor} =\\\text{strength in}\\\text{table/strength}\\\text{of chemical}\\\text{selected}\end{array}}} \div \underset{4}{\boxed{\begin{array}{c}\text{Volume}\\\text{(from}\\\text{Table 5-3)}\end{array}}} = \begin{array}{c}\text{Amount}\\\text{of}\\\text{chemical}\\\text{needed}\end{array} \quad \text{(c5-5)}
$$

This calculator (c5-5) allows for the use of any chemical, or any strength, for any size storage facility. Just plug in the numbers and calculate the answer. Make sure that all of the values are either in US or SI units.

- The first value is the volume of the tank. This comes from the tank volume calculators or Table 5-2 for horizontal cylindrical tanks.
- Next, insert the value from Table 5-3 for the chemical selected (liquid chlorine, sodium hypochlorite, or calcium hypochlorite).
- The third factor is an adjustment for the strength of the chemical if it differs from the amount in Table 5-3. The

factor is the strength shown in the table divided by the strength you are using.

- The fourth factor is the volume used in Table 5-3 (either 100,000 gallons for US units or 100 cubic meters for SI units).

Example 1.   The full storage facility chlorination method is used for a vertical cylindrical tank that is 20 ft in diameter and the maximum depth to the overflow is 30 ft. The disinfection chemical selected is 10 percent sodium hypochlorite. Use all US units.

1. Volume from vertical cylinder calculator

   5.87 × 20 ft × 20 ft × 30 ft = 70,400 gal

2. Amount of chemical from Table 5-3 for 5% sodium hypochlorite

   19.4 gal

3. Chemical strength factor is Table 5-3 strength/strength in example

   5% ÷ 10% = 0.5

4. Volume used in Table 5-3

   100,000 gal

Plug these values into the calculator.

   70,400 × 19.4 × 0.5 ÷ 100,000 = 6.8 gal of 10% sodium hypochlorite

---

Example 2.   Using the full storage facility chlorination method for a 10 mg/L chlorine dosage and a 24 hr contact time, how much 65 percent calcium hypochlorite is needed for a rectangular tank that is 10 m long, 5 m wide, and has a maximum water depth of 5 m? Use all SI units.

1. Volume from rectangular tank calculator

   10 × 5 × 5 = 250 m³

2. Amount of chemical from Table 5-3 for 65% calcium hypochlorite

   1.5 kg
3. Chemical strength factor is Table 5-3 strength/strength in example

   65% calcium hypochlorite ÷ 65% calcium hypochlorite = 1
4. Volume used in Table 5-3

   100 m³

Plug these values into the calculator.

   250 × 1.5 × 1 ÷ 100 = 3.75 kg of 65% calcium hypochlorite

Surface Application Method

In this method, a 200 mg/L chlorine solution is applied to all surfaces within the storage facility that will be in contact with the water when the facility is completely full.

Chlorination Steps and Precautions

1. Apply the chlorine solution to all exposed surfaces (Table 5-4) with suitable brushes or spray equipment (Figure 5-2). Portable spray equipment is adequate for small tanks.
2. Add chlorine to drains or drain piping so that the chlorine residual of the pipe when filled with water shall be at least 10 mg/L. Overflow piping does not need to be chlorinated. Make sure the drain line valve is closed to avoid the need for dechlorination.
3. Allow at least 30 min for contact of the strong chlorine solution with the treated surfaces.
4. Introduce potable water and remove the water from treated drains and drain piping.
5. Fill the facility to overflow level with potable water and return to service.

Table 5-4 Approximate chlorination chemical requirements to prepare 200 mg/L chlorine solutions

Volume of Water		Sodium Hypochlorite Solution Requirements								Calcium Hypochlorite 65% Available Chlorine	
		1% Strength		5% Strength		10% Strength		15% Strength			
gal	L	cup*	L	cup	L	cup	L	cup	L	oz†	g
5	18.95	1.6	0.39	0.32	0.075	0.16	0.04	0.11	0.025	0.2	6
10	37.9	3.2	0.78	0.64	0.15	0.32	0.08	0.22	0.05	0.4	12
20	75.8	6.4	1.56	1.28	0.3	0.64	0.16	0.44	0.1	0.8	23
50	189.3	16 (1 gal)	3.9	3.2	0.75	1.6	0.4	1.1	0.25	2.1	58
100	378.5	32 (2 gal)	7.8	6.4	1.5	3.2	0.8	2.2	0.5	4.1	116
150	567.8	48 (3 gal)	11.7	9.6	2.25	4.8	1.2	3.3	0.75	6.2	175
200	757	64 (4 gal)	15.6	12.8	3.0	6.4	1.6	4.4	1.0	8.2	233

* A cup is 8 fl. oz.
† This is an ounce of weight. There are 16 oz in a pound.

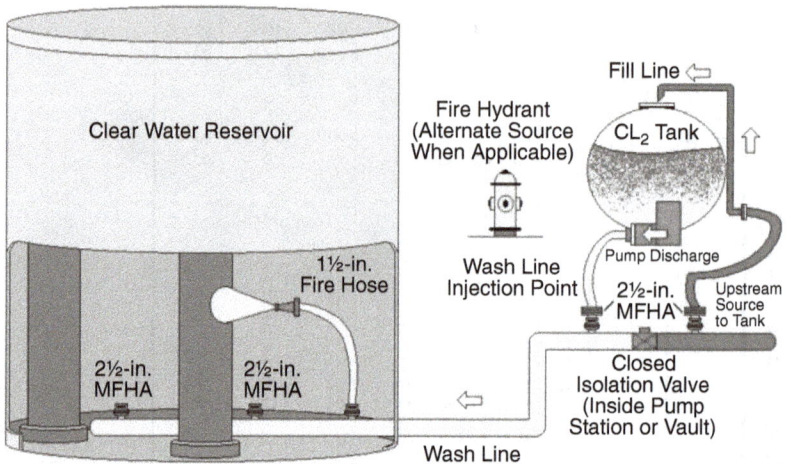

Courtesy Nicole Peschel
Figure 5-2   Surface application method installation (large tank example)

Chlorination Amount Calculation

This method uses a 200 mg/L chlorine solution (prepared according to Table 5-4) to spray all of the exposed surfaces inside the storage facility. The volume needed will vary depending on the size of the facility. Sodium and calcium hypochlorite are commonly used to make the chlorine solution used for this method.

> **Example 1.** How much 10% sodium hypochlorite is needed to prepare 150 gal of 200 mg/L chlorine solution needed to spray the inside of a storage facility?
> From Table 5-4: *4.8 cups or 1.2 L are needed.*

Chlorinate and Fill Method

Chlorine and water are added so that about 5 percent of the storage volume of the facility is filled (Figures 5-3, 5-4, and 5-5), and this water has a chlorine residual of approximately 50 mg/L. After a retention period of at least 6 hr, the facility is filled to

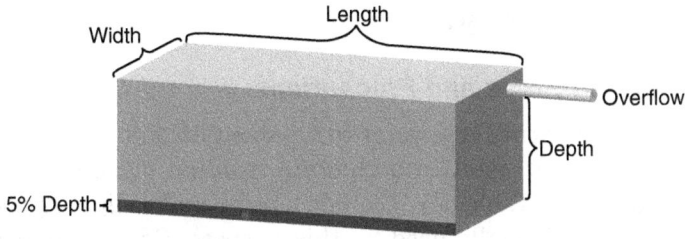

Figure 5-3   Rectangular or square storage tank

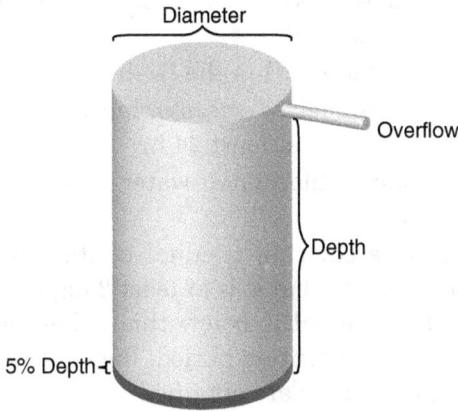

Figure 5-4   Vertical (upright) cylindrical storage tank

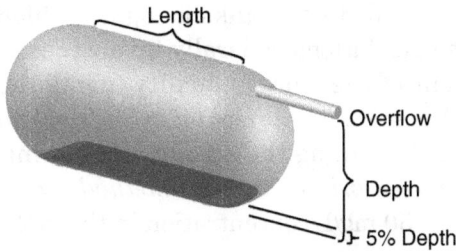

Figure 5-5   Horizontal cylindrical storage tank (with hemispheres on both ends)

overflow with potable water. The resultant chlorine residual is suitable for potable use.

Chlorination Steps and Precautions

1. Add chlorine and water to fill about 5% of the storage volume and result in a chlorine residual of approximately 50 mg/L (Table 5-5). Liquid chlorine and sodium hypochlorite are usually used for this purpose. Calcium hypochlorite granules or tablets can also be used provided that adequate mixing is available to ensure that the chemical fully dissolves and is distributed evenly throughout the facility. Follow procedures described for the full facility chlorination method.
2. Retain this water within the facility for at least 6 hr.
3. Fill the remainder of the storage facility with potable water and hold for at least 24 hr.
4. Remove highly chlorinated water from drains and drain piping.
5. Test the free chlorine residual of the water in the full facility to ensure that it is at least 2 mg/L. Add more chlorine if the residual is below this value and retest after another 24 hr retention period. Repeat this process until satisfactory results are obtained.

Chlorination Amount Calculation

The calculators (c5-1, 5-2, 5-3, 5-4, 5-5, 5-6) and Tables 5-5 and 5-6 should be used to calculate the amount of chlorination chemical needed to disinfect tanks using the Chlorinate and Fill Method. These calculations generally involve calculating the volume of 5 percent of the storage facility (using the *tank volume calculators or Table 5-2 for horizontal cylindrical tanks and Table 5-5*) and then determining the amount of chlorination chemical needed (*using the chlorinate and fill method chemical calculator c5-6*) to achieve a 50 mg/L concentration in that volume.

Table 5-5   Approximate depth to obtain 5% of total volume

Tank Shape	5% Volume Depth
Rectangular, square, vertical cylindrical	0.05 × maximum depth to overflow
Horizontal cylindrical	0.1 × maximum depth to overflow (slightly less)

Table 5-6   Approximate amount of chlorination chemicals needed for 50 mg/L concentration

Volume Treated	Liquid Chlorine		5% Sodium Hypochlorite		65% Calcium Hypochlorite	
	lb	kg	gal	L	lb	kg
100,000 gal	42	19	97	367	64	29
100 m³	11.1	5.0	25.6	97.0	16.9	7.7

Chlorinate and Fill Method Chemical Calculator

The approximate amount of chlorination chemical needed for 50 mg/L chlorine concentration for a given size tank, 5 percent volume, is calculated as follows (use either US or SI units):

1		2		3		4		
Volume (from tank shape equations × 0.05)	×	Amount of chemical (from Table 5-6)	×	Chemical strength factor = strength in table/strength of chemical selected	÷	Volume (from Table 5-6; 100,000 for US units and 100 for SI units)	=	Amount of chemical needed (c5-6)

This calculator (c5-6) is for the chlorinate and fill method and, similar to the one on page 74, it allows for the use of any chemical, or any strength, for any size storage facility. Make sure that all of the values are either in US or SI units.

- The first value is 5% of the volume of the tank. This comes from the tank volume calculators or Table 5-2 for horizontal cylindrical tanks and then multiplied by 0.05 to obtain the 5% volume (for the horizontal cylindrical tank use 0.1 as the factor to obtain the 5% volume).
- Next, insert the value from Table 5-6 for the chemical selected (liquid chlorine, sodium hypochlorite, or calcium hypochlorite).
- The third factor is an adjustment for the strength of the chemical if it differs from the amount in Table 5-6. The factor is the strength shown in the table divided by the strength you are using.
- The fourth factor is the volume used in Table 5-6 (either 100,000 gallons for US units or 100 cubic meters for SI units).

Example 1.   The chlorinate and fill method is used for a vertical cylindrical tank that is 20 ft in diameter and the maximum depth to the overflow is 30 ft. The disinfection chemical selected is 10% sodium hypochlorite.
Use all US units.

1. Volume from vertical cylindrical tank calculator
   5.87 × 20 ft × 20 ft × 30 ft × 0.05 = 70,400 gal × 0.05 = 3,520 (for 5% depth and volume)
2. Amount of sodium hypochlorite from Table 5-6
   97 gal
3. Chemical strength factor is Table 5-6 strength/strength in example
   5% ÷ 10% = 0.5
4. Volume used in Table 5-6
   100,000 gal

Plug these values into the calculator.
   3,520 × 97 × 0.5 ÷ 100,000 = 1.7 gal of 10% sodium hypochlorite

---

Example 2. Use the chlorinate and fill chlorination method for a 50 mg/L chlorine dosage and a 24-hr contact time. How much 65 percent calcium hypochlorite is needed for a rectangular tank that is 10 m long, 5 m wide, and has a maximum water depth of 5 m?

Use all SI units.

1. Volume from rectangular tank calculator
   $$10 \times 5 \times 5 \times 0.05 = 250 \ m^3 \times 0.05 = 12.5 \ m^3$$
2. Amount of calcium hypochlorite from Table 5-6
   7.7 kg
3. Chemical strength factor is Table 5-6 strength/strength in example
   65% calcium hypochlorite ÷ 65% calcium hypochlorite = 1
4. Volume used in Table 5-6
   100 m³

Plug these values into the calculator.

$$12.5 \times 7.7 \times 1 \div 100 = 0.96 \ kg \ of \ 65\% \ calcium \ hypochlorite$$

References

ANSI/AWWA C652. *AWWA Standard for Disinfection of Water-Storage Facilities.* American Water Works Association: Denver, Colo.

Standard Methods for the Examination of Water and Wastewater (current edition). APHA, AWWA, WEF: Washington, D.C.

Chapter 6

Underwater Inspection of Storage Facilities—Disinfection Procedures

Storage facilities can be inspected by divers or remote operated vehicles (ROV) while the facility is filled with potable water. This procedure must be conducted by trained and certified personnel. Great care should be used to ensure that the potable water contained in the facility is not contaminated. The procedures described here and in *ANSI/AWWA C652 Standard for Disinfection of Water-Storage Facilities* include the minimum requirements. Contractors and owners may employ additional procedures to increase the safety and effectiveness of these procedures. Underwater inspection of a potable water storage facility may be prohibited by regulatory agencies in some locations. Applicable regulatory agencies should be contacted to determine their requirements before beginning this procedure.

In most cases, underwater inspection is performed by contractors to the utility. Therefore, this chapter is written as if a contractor is employed for this purpose. Some utilities have the trained personnel and equipment to perform this procedure. In this case, the utility employees would act as a contractor and the reader can make that substitution where it is appropriate.

Preliminary Meeting
A meeting should be held involving all parties to ensure that everyone understands the issues and that clear lines of communication are established. Some topics to discuss include:

- Configuration of the reservoir
- Disinfection procedures
- Identification of appurtenances that may be encountered
- Timing and time restrictions

- Diving conditions
- Safety procedures
- Inspection requirements
- Equipment delivery and acceptance

Storage-Facility Isolation

Underwater inspection can be performed safely whether or not the facility is isolated from the distribution system. In either case, safety procedures must be followed. The facility should be as full as possible while allowing space for safe entry and exit. If the facility cannot be isolated from use, underwater inspection can still be performed, but operations should ensure that water is entering the facility rather than exiting during the time of the inspection. A constant and minimum flow rate should be maintained for the duration of the inspection procedure. This will help provide a stable and safe environment.

Storage-Facility Access

The area around the access hatch should be cleaned before opening. Previously disinfected equipment must be redisinfected if it contacts the roof or other areas that have not been disinfected when entering the storage facility.

Initial Water Quality

The water in the facility should be sampled before entering (and before placing any equipment within the facility). Several samples should be taken from various locations to establish the preinspection water quality. The test results should be recorded for chlorine residual and turbidity (total coliform tests are also recommended) for comparison purposes at the conclusion of the inspection procedure.

Equipment and Personnel Requirements

Only dedicated equipment and clothing should be used for underwater storage facility inspection. Equipment materials contacting the potable water should not adversely affect the water quality. Both self-contained underwater breathing apparatus

(SCUBA) and externally supplied air sources are acceptable for use in this procedure. Disinfection should not damage any equipment or materials exposed to the water. Divers must be completely encapsulated in a dry-suit (dedicated for this purpose) with the head enclosed by a helmet or dry-suit hood with full-face mask. Diver communication capability must comply with applicable regulations.

When inspecting the storage facility with divers or ROVs, care should be taken to avoid disturbing sediment on the bottom (floor) of the facility. Resuspended sediments may affect the water quality by releasing bacteria and increasing turbidity.

Diving personnel or other persons involved in storage facility underwater dive operations must comply with all applicable safety requirements and regulations. Qualified divers should provide documentation of direct tank inspection and cleaning work experience. Acceptable qualifications include but are not limited to the following:

- 2nd Class US Navy Diver Training
- ANSI/ACDE 01 Commercial Diver Certification
- ADC Commercial Certification

All personnel entering the storage facility must have OSHA Confined Space Certification (if required). Certificates must be available on-site. These personnel must be free of communicable disease (and not under a physician's care within seven days prior to the dive). All dive and appropriate support personnel shall be certified for CPR and first aid by the American Red Cross (or equivalent agency). A comprehensive safety manual must be on-site during the procedure. Certifications for all personnel involved should be in the manual. Emergency extraction equipment (or availability of properly trained and equipped rescue squad) must be provided.

Equipment Disinfection

ANSI/AWWA Standard 652 requires the disinfection of all equipment immediately prior to entering the storage facility. Submerge, spray (Figure 6-1), or swab a 200 mg/L chlorine solution (Table 5-3) on all surfaces including diver covering (dry-suit,

Figure 6-1   Diver disinfection

breathing equipment, face mask, etc.) according to the *surface application method* described in Chapter 5.

Postinspection Chlorine Residual

Retest the chlorine residual following the inspection procedure. The chlorine residual value should match the value obtained before the procedure. Chlorine may be added to the facility (using calcium hypochlorite granules or sodium hypochlorite solution to the water surface) to bring the residual to the preinspection level. Mixing by circulation (predisinfected mixers or pumps may be used) should be provided.

Bacteriological Testing

Sampling should be conducted and testing provided as described in Chapter 5. When results are satisfactory, the storage facility can be returned to service. Unsatisfactory results require retesting and possible redisinfection (Chapter 5).

Affidavit of Compliance

Contractors may be required to provide documentation that attests to compliance with all aspects of the disinfection standard (ANSI/AWWA C652).

References

ANSI/AWWA C652. *AWWA Standard for Disinfection of Water-Storage Facilities.* American Water Works Association: Denver, Colo.

Chlorine Institute. 1995. *Pamphlet 100, Sodium Hypochlorite Handling and Storage Guidelines.* Washington D.C.

Chlorine Institute. 1998. *Communicating Your Chlorine Risk Management Plan.* Washington, D.C.

Chlorine Institute. 1998. *Pamphlet 90, Molecular Chlorine: Health and Environmental Effects.* Washington, D.C.

Chlorine Institute. 1999. *Water and Wastewater Operators Chlorine Handbook*, 1st ed. Washington, D.C.

Chlorine Institute. 2000. *The Chlorine Manual*, 6th ed. Washington, D.C.

Chlorine Institute. 2001. *Pamphlet 65, Personal Protective Equipment for Chlor-Alkali Chemicals.* Washington, D.C.

Standard Methods for the Examination of Water and Wastewater (current edition). APHA, AWWA, WEF: Washington, D.C.

Chapter 7

Dechlorination of Highly Chlorinated Water

Dechlorination is the practice of partially or totally removing the chlorine residual. Highly chlorinated water used for the disinfection of facilities, pipelines, and appurtenances is commonly dechlorinated when discharged. In some areas, low chlorine residual drinking water must be dechlorinated before it can be drained to storm sewers or directed to waterways. Water discharged during pipeline break emergencies must occasionally be dechlorinated.

Dechlorination practices generally have not been optimized to the degree of chlorination operations. Local conditions and regulations have driven the need to dechlorinate and the development of methods that address these situations. Some of the most important factors that need to be considered are: the flow (or volume) to be dechlorinated; the chlorine residual level (initial and final); the time (or distance) that dechlorination must achieve; the location and the chemistry of the water to be dechlorinated; the impact on the receiving waterway or facility; and any regulatory requirements. System operators should evaluate all of the applicable factors and select the dechlorination method that best suits their situation.

The information provided in this chapter reflects the array of practices currently used by water utilities to dechlorinate water in the field. These are not the only methods but are the most common ones. In some cases, the performance values presented are from empirical field observations and may not predict the performance of a practice used in another situation. Operators should test any dechlorination procedure to determine the exact conditions for that application.

Dechlorination Practices Used in the Field

There are a number of nonchemical and chemical dechlorination methods. The selection of a method depends on site-specific factors, such as cost, availability of chemicals, containment logistics, availability of specialized equipment, and regulatory approval.

Nonchemical techniques include retention in holding ponds, land application, groundwater recharge, discharge through hay bales and other natural obstructions, and discharge into sanitary sewers. These methods have the advantage of simplicity because they avoid the issues connected with the storage, handling, and safety concerns related to dechlorination chemicals.

Chemicals commonly used for dechorination in the field include sodium bisulfite, sodium metabisulfite, sodium sulfite, sodium thiosulfate, calcium thiosulfate, ascorbic acid, and sodium ascorbate. Sulfur dioxide is a common dechlorination chemical, but it is not often used in field applications, and its use is not included in this discussion. Chemicals have advantages over nonchemical methods because they usually require less time to affect dechlorination.

Some of the dechlorination chemicals pose potential health concerns if not handled properly and may cause adverse environmental impacts. For example, sodium bisulfite and sodium metabisulfite are skin, eye, or respiratory tract irritants. Sulfite-based chemicals can cause water quality concerns by depleting dissolved oxygen in receiving streams. Some dechlorination chemicals produce hydrochloric acid and therefore, decrease water pH. When selecting a chemical for dechlorination, it is important to consider the by-products of the reaction and to receive approval from the appropriate regulatory agency.

Treat and Test Field Method

Although the amount of chemical needed for dechlorination (and to some degree, the effectiveness of nonchemical methods) can be calculated as indicated in this chapter, as a practical matter, most operators use the "treat and test" method of dechlorination in the field. The operator chooses a method for dechlorination (nonchemical or chemical) and then measures the result. If the desired result is not obtained, the method is adjusted until it is. This is a trial-and-error method.

In some cases, a combination of methods is used to provide backup while the system is adjusted. Frequent samples are taken and tested to ensure that the dechlorination is successful. If a chemical method is used, the feed amount is adjusted so that the desired chlorine residual is just attained. This is necessary to avoid an overfeed situation. Overfeeding some dechlorination chemicals can have adverse effects on receiving waters.

Calculating the amount of dechlorination chemical necessary for the volume and concentration present is recommended so that enough chemical is on-hand to complete the job. The tables and equations in this chapter are provided to perform these calculations. It is also recommended that some extra chemical (at least 20 percent) is taken to the site, because some of the calculations are approximate and local conditions may affect the exact amount needed.

Frequent chlorine residual testing is needed when employing this method. Use reliable, certified test methods and field test kits for this purpose. Personnel must be trained in the proper use of this test equipment. Regulatory agencies may require documentation of the test results and other applicable information such as the initial volume, the chlorine residual, the dechlorination method used, and the amount of dechlorination chemical.

Nonchemical Methods

These dechlorination methods are generally used for discharges where the amount of water is limited. The reason for this is that most of these methods require some time to affect the chlorine reduction. They are not generally suitable for continuous flow treatment. The most common nonchemical methods and estimates of dechlorination effectiveness are given in Table 7-1.

Chemical Dechlorination Methods

Common reducing agents used in the field include sodium bisulfite ($NaHSO_3$), sodium sulfite (Na_2SO_3), sodium metabisulfite ($Na_2S_2O_5$), sodium thiosulfate ($Na_2S_2O_3$), calcium thiosulfate (CaS_2O_3), ascorbic acid (Vitamin C), and sodium

Table 7-1   Nonchemical dechlorination methods

Method of Discharge	Dechlorination Effectiveness Notes
Retention in holding tanks	Chlorine (0.5–2.0 mg/L) reduced in several hours to a few days. Chloramines take three to four times longer.
Land application/flow over pavement or gravel	Chlorine (1.2 mg/L) at 300 gpm reduced to 1.0 mg/L in 500 ft. Chloramines (1–2 mg/L) at 300 gpm needed 2,414 ft to reduce the residual below regulatory requirements.
Groundwater recharge/ percolation into groundwater basins	Very effective but groundwater must be carefully monitored. Only approved in limited areas.
Through natural obstructions (hay bales)	May be effective but control is difficult. A crude technique that has limited application.
Storm sewers	Chlorine demand may not be high enough to eliminate a residual prior to discharge to the receiving waterway. May be acceptable depending on the circumstances.
Sanitary sewers	The method most used by water utilities. Must have a sewer entry point near the location of discharge. Need authorization from the sewage agency. Backflow prevention must be practiced.

ascorbate ($NaC_6H_7O_6$). A comparison of these chemical dechlorination agents is provided in Table 7-2. The amount of dechlorination chemical needed for water with a 100 mg/L chlorine residual for each 100 ft (30 m) of pipeline is given in Tables 7-3 and 7-4.

Chemical Feed Techniques

There are several methods commonly used for feeding dechlorination chemicals in the field. Some of these may require the availability of electricity. Portable generators can provide electricity in areas where it may not be readily available.

Gravity feed method. The gravity feed method typically involves adding dechlorinating solution from a container equipped with a spigot that is placed above the water flow path.

Table 7-2   Comparison of dechlorination agents

Agent	Forms Available	Dose at pH 7.0 mg/mg Cl*	pH of 1% Solution	Decomposition/ Off-gassing	Oxygen Scavenger
Sodium bisulfite	Powder/ crystal	1.61	4.3	Sulfur dioxide	Strong
Sodium metabisulfite	Powder/ crystal	1.47	4.3	Sulfur dioxide	Strong
Sodium sulfite	Powder/ crystal/ tablet	1.96	8.5–10.0	Sulfur dioxide	Moderate
Sodium thiosulfate	Powder/ crystal	1.9	7.0	Sulfur dioxide	Weak
Calcium thiosulfate	Powder/ crystal	1.22	6.5–7.0	Sulfur dioxide	Weak
Ascorbic acid	Powder/ crystal/ tablet	2.5	6.0*	none	None reported
Sodium ascorbate	Crystal/ tablet	2.8	7.0	none	None reported

* All dechlorination dosage estimates listed are based on field results (values may differ based on variable conditions). Dechlorination capacity must be checked for each application using site-specific conditions. The symbol Cl in this heading refers to chlorine as hypochlorous acid (HOCl).

Table Tamer   This table shows the approximate dosage of various dechlorination chemicals needed to neutralize one milligram of chlorine. Example: About 1.61 mg of sodium bisulfite is needed for each milligram of chlorine.

The discharge spigot on the container can be adjusted to provide a minimum dechlorinating solution feed rate into the water flow, based on calculations involving the concentration of the dechlorinating solution, water flow rate, and residual chlorine concentration in the flow stream.

To minimize the volumes of dechlorinating solution needed on field vehicles, wherever possible, dry dechlorinating agent should

Table 7-3   Approximate amount of dechlorination chemical needed for 100 mg/L chlorine residual and 100 ft of pipeline (US units)

Pipeline Diameter in.	Pounds of Chemical Needed							
	Sodium Thiosulfate	Sodium Bisulfite	Sodium Metabisulfite	Sodium Sulfite	Calcium Thiosulfate	Ascorbic Acid	Sodium Ascorbate
2	0.03	0.02	0.02	0.03	0.02	0.03	0.04
4	0.10	0.09	0.08	0.11	0.07	0.14	0.15
6	0.23	0.20	0.18	0.24	0.15	0.31	0.34
8	0.41	0.35	0.32	0.43	0.27	0.54	0.61
10	0.65	0.55	0.50	0.67	0.41	0.85	0.95
12	0.93	0.79	0.72	0.96	0.60	1.22	1.37
16	1.65	1.40	1.28	1.71	1.06	2.18	2.44
18	2.09	1.77	1.62	2.16	1.34	2.75	3.08
20	2.58	2.19	2.00	2.67	1.66	3.40	3.81
24	3.72	3.15	2.88	3.84	2.39	4.90	5.48
30	5.81	4.93	4.50	6.00	3.73	7.65	8.57
36	8.37	7.09	6.48	8.64	5.38	11.01	12.34
42	11.39	9.65	8.82	11.75	7.32	14.99	16.79
48	14.88	12.61	11.51	15.35	9.56	19.58	21.93
54	18.83	15.96	14.57	19.43	12.09	24.78	27.76
60	23.25	19.70	17.99	23.99	14.93	30.60	34.27
64	26.46	22.42	20.47	27.29	16.99	34.81	38.99

Table 7-4   Approximate amount of dechlorination chemical needed for 100 mg/L chlorine residual and 30 m of pipeline (SI units)

Pipeline Diameter mm	Grams of Chemical Needed						
	Sodium Thiosulfate	Sodium Bisulfite	Sodium Metabisulfite	Sodium Sulfite	Calcium Thiosulfate	Ascorbic Acid	Sodium Ascorbate
50	11.19	9.48	8.65	11.54	7.18	14.72	16.49
100	44.75	37.92	34.62	46.16	28.73	58.88	65.94
150	100.68	85.31	77.89	103.86	64.64	132.47	148.37
200	178.98	151.66	138.47	184.63	114.92	235.50	263.76
250	279.66	236.97	216.37	288.49	179.57	367.97	412.13
300	402.71	341.24	311.57	415.42	258.58	529.88	593.46
400	715.92	606.65	553.90	738.53	459.70	942.00	1,055.04
450	906.09	767.79	701.02	934.70	581.80	11,92.22	1,335.29
500	1,118.63	947.89	865.46	1,153.95	718.28	1,471.88	1,648.50
600	1,610.82	1,364.96	1,246.27	1,661.69	1,034.32	2,119.50	2,373.84
800	2,863.68	2,426.59	2,215.58	2,954.11	1,838.78	3,768.00	4,220.16
900	3,624.35	3,071.16	2,804.10	3,738.80	2,327.21	4,768.88	5,341.14
1,000	4,474.50	3,791.55	3,461.85	4,615.80	2,873.10	5,887.50	6,594.00
1,200	6,443.28	5,459.83	4,985.06	6,646.75	4,137.26	8,478.00	9,495.36
1,400	8,770.02	7,431.44	6,785.23	9,046.97	5,631.28	11,539.50	12,924.24
1,500	10,067.63	8,530.99	7,789.16	10,385.55	6,464.48	13,246.88	14,836.50
1,600	11,454.72	9,706.37	8,862.34	11,816.45	7,355.14	15,072.00	16,880.64

be mixed directly within the container prior to use, rather than using premixed dechlorinating solutions provided by suppliers. As part of the dechlorination procedure, samples can be collected downstream of the feed point, analyzed for pH, dissolved oxygen, and residual chlorine, and if necessary, the chemical feed rate can be adjusted to ensure a nondetectable residual chlorine concentration prior to discharge.

Gravity feed systems are simple to operate, have minimal equipment requirements, have been used effectively by various utilities, and are inexpensive (low-density polyethylene carboys equipped with spigots). Unless adjusted, chemical feed rates can be expected to decrease slightly over time, however, as the available head within the carboy decreases during use. A disadvantage of using this technique is that it involves field testing and calculations for flow rates and water quality parameters to adjust chemical feed rate. Field maintenance crews sometimes prefer a method that does not involve field calculations.

Chemical metering pumps. This method for injecting a dechlorinating solution is similar to the gravity feed method except that a chemical metering pump is used to inject the dechlorinating solution from a container into the water flow. Chemical metering pumps are capable of delivering chemical solutions over a wide range of flows, the flow rates (Figure 7-1) are adjustable, and the pumps provide a constant chemical feed rate.

Relative to a gravity feed system, this type of system requires more equipment (e.g., storage container, pump, energy source, and tubing), and costs are significantly higher. Although chemical metering pumps may provide field personnel with a more reproducible method, this feature involves much higher cost and greater operator ability and attention during dechlorination.

Venturi injector systems. Venturi injectors are differential pressure injection devices that allow for the injection of liquids (e.g., dechlorinating solutions from hydrants, etc.) into a pressurized water stream. This method is well suited for pressurized water releases such as those through hydrants.

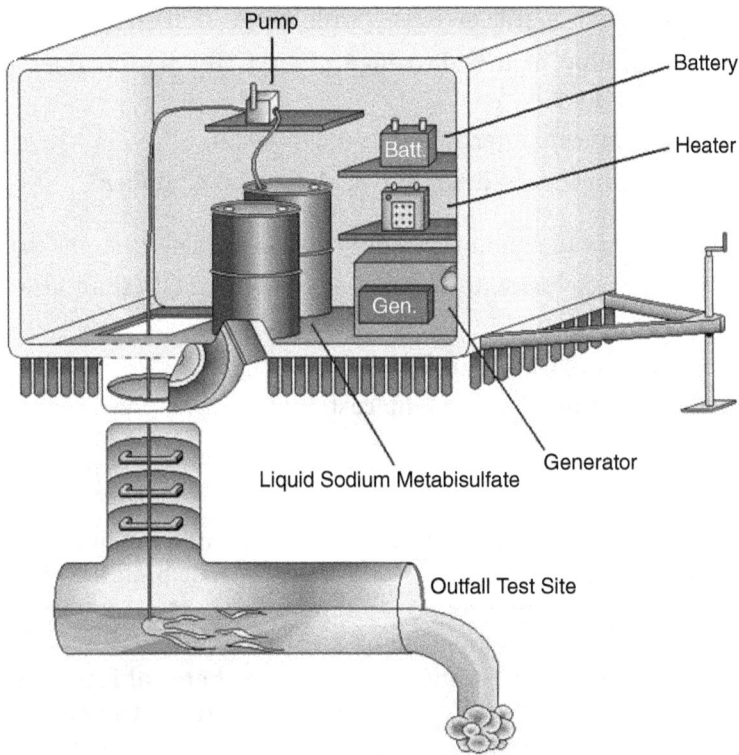

Courtesy Nicole Peschel

Figure 7-1   Portable dechlorination metering installation

Water main discharges are constricted by a regulator valve or gate valve and routed through a Venturi injector system. Dechlorinating solution is drawn into the injector from a plastic container, with the chemical feed rate controlled by a metering valve. Venturi injector units should include a flowmeter installed near the metering valve to measure the chemical feed rate (e.g., a rotameter); a threaded fitting at the upstream end of the pipe to attach adapters (e.g., reducing sections); and a fitting at the downstream end of the pipe for attachment of flexible discharge piping (e.g., a hose).

The primary advantages of Venturi injection systems are

- efficient operation over a wide range of pressures;
- available in a wide range of sizes, flows, and injection capacities;
- no external energy requirements; and
- instantaneous mixing in the injection chamber.

The primary disadvantages of Venturi injection systems are

- more sophisticated equipment requirements than gravity feed systems;
- slightly more labor intensive set-up;
- higher unit equipment cost than gravity feed systems; and
- require constant monitoring.

Spray feed systems.   The dechlorinating solution can be sprayed into the flow (on pipe walls, surfaces, or pipeline appurtenances) via a backpack sprayer similar to those used for pesticide and herbicide application or a chemical fire extinguisher. The advantage of this technique is that the chemical feed rate is fairly constant (given a steady pressure within the solution chamber). Therefore, dosages can be approximated fairly accurately, and a piped or channelized flow is not required to effectively feed the chemical. This method is typically more effective in adding dechlorinating chemicals to sheet flows than the other alternatives previously described.

A significant disadvantage of a spray feed system is that it requires the equipment to be set up at a stationary point and monitored for adequate chemical and pressure. If a stationary point is not available, a person will have to don the sprayer and continuously apply the dechlorinating agent to the flow.

Flow-through systems.   Flow-through systems (Figure 7-2) include any method where the solid chemical is held stationary and the flow is allowed to run over, around, or through it. Examples include pumping chlorinated water through a container filled with dechlorinating agent, or laying permeable bags of the

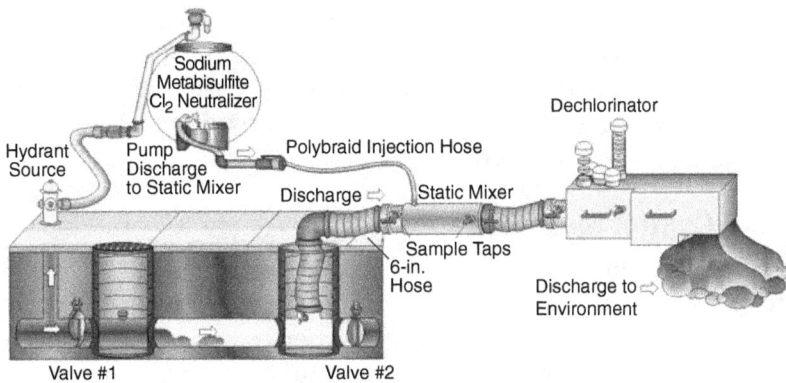

Courtesy Nicole Peschel
Figure 7-2   Pipeline dechlorination system

chemical in the flow path. For this application, the dechlorinating agent is in tablet or powder form.

The advantages of flow-through systems are that they are simple and can be used for sheet flow applications as well as for channelized or pumped flow. The disadvantages are that there is limited control over the dosage; overdosing or underdosing to significant levels could occur; and it may be difficult, in some cases, to tell when the chemical has been used up and must be replaced.

Also, because of the contact time required for dissolution of powder/tablets, this method is more suitable for low- and medium-velocity discharges. This method may, however, be suitable for dechlorinating releases from unidirectional flushing where the velocity of the flow is approximately 5.0 ft/sec provided that equipment is specifically designed for this flow.

A variation in the application of a flow-through system is the automatic tablet dispenser. These devices are similar to the tablet feeders currently used for disinfection in many water/wastewater treatment plants. The feeders typically consist of a nonmoving housing and automatic feed tubes. The tubes are inserted down through a removable top cover of the feeder into the stream of water. The lower end of each tube is slotted to permit free flow of water through the tubes to assure good contact between the water and dechlorination tablets. As the stream of water flows past the feed tubes containing dechlorination tablets, the dechlorination agent is released into the water by dissolution. Commercial tablet dispensers are available for a wide variety of flow rates and chlorine concentrations.

Flow Control Measures

During planned and unplanned water releases, it may be necessary to construct flow control measures to prevent the water from entering directly into a water body and to provide an opportunity for better mixing of the dechlorinating agent. Construction of berms, swales, ditches, or redirection pipes are common methods used to control the flow of released water.

Chemical Dechlorination Calculation

Table 7-5 lists the amount of dechlorination chemical needed to remove 1 mg/L residual for 100,000 gal or 100 m³. The volume of water is calculated from Table 3-6 for pipelines, and Table 5-2 and calculators c5-1, 5-2, 5-3, and 5-4 for storage tanks. Make sure that the volumes are in gallons for US units and cubic meters for SI units. The total amount of dechlorination chemical needed is calculated using the *dechlorination chemical calculator.*

Table 7-5   Approximate dechlorination chemical needed for fixed volume

Pounds of Chemical Needed for 100,000 gal per 1 mg/L						
Sodium Thiosulfate	Sodium Bisulfite	Sodium Metabisulfite	Sodium Sulfite	Calcium Thiosulfate	Ascorbic Acid	Sodium Ascorbate
1.58	1.34	1.23	1.63	1.02	2.08	2.33

Kilograms of Chemical Needed for 100 m³ per 1 mg/L						
Sodium Thiosulfate	Sodium Bisulfite	Sodium Metabisulfite	Sodium Sulfite	Calcium Thiosulfate	Ascorbic Acid	Sodium Ascorbate
0.19	0.16	0.15	0.20	0.12	0.25	0.28

Dechlorination Fixed Volume Chemical Calculator

1		2		3		4		(c7-1)
Volume of water to be dechlorinated (from Tables 3-6 and 5-2, and calculators c5-1, 5-2, 5-3, and 5-4)	×	Amount of chemical (from Table 7-5)	×	Chlorine residual, mg/L	÷	Volume (from Table 7-5, either 100,000 gal or 100 m³)	=	Total amount of chemical needed

This calculator is used for calculating the amount of dechlorination chemical for a fixed volume and it has four factors. Insert the values for each factor and calculate the amount of chemical needed for this procedure.

- The first factor is the total volume of water to be dechlorinated. This is the volume of the pipeline, slug, or storage tank chlorination water. The amount is obtained from the applicable table (3-6, 5-2) or calculator (c5-1, 2, 3, 4).
- The second factor is the amount of chemical from Table 7-5.
- Next is the chlorine residual in mg/L.
- Last is the volume in Table 7-5 (either 100,000 gallons or 100 cubic meters).

Example 1.   Calculate the amount of sodium sulfite needed to dechlorinate a chlorine concentration of 10 mg/L in a vertical cylindrical tank that is 20 ft in diameter and is 35 ft maximum depth.

1. Using calculator c5-4: $5.87 \times 20 \times 20 \times 35 = 82{,}180$ gal
2. 1.63 lb
3. 10 mg/L
4. 100,000 gal

   $82{,}180 \times 1.63 \times 10 \div 100{,}000 = 13.43$ lb of sodium sulfite

---

Ops Tip   Insert values in calculator c7-1 above

Ops Tip   Insert values in calculator c7-1 above

> **Example 2.**   Calculate the amount of ascorbic acid needed to dechlorinate the water from 500 m of a 200-mm diameter pipeline filled with 25 mg/L chlorine solution.
>
>   1. Using Table 3-6: 31.4 L/m × 500 m = 15,700 L or 15.7 m^3
>   2. 0.25 kg
>   3. 25 mg/L
>   4. 100 gal
>   *15.7 × 0.25 × 25 ÷ 100 = 0.98 kg of ascorbic acid*

Dechlorination Continuous Solution Feed Chemical Calculator

A 1% solution of any of the dechlorination chemicals listed in Table 7-2 (assuming 100% strength solid to start) is prepared by:
1. Mixing 0.083 lb of solid with each gallon of water
2. Mixing 1.3 oz of solid with each gallon of water
3. Mixing 10 g of solid with each liter of water
4. Mixing 10 kg of solid with each cubic meter of water

(c7-2)

Dechlori-nation chemical feed rate (gph)	= 0.006 ×	Chlorinated water flow rate (gpm) (withdrawal or fill)	×	Chlorine residual (mg/L)	×	Dechlori-nation chemical dosage (Table 7-2)	÷	Strength of dechlori-nation chemical solution (%)

(c7-3)

Dechlori-nation chemical feed rate (mL/min)	= 0.038 ×	Chlorinated water flow rate (gpm) (withdrawal or fill)	×	Chlorine residual (mg/L)	×	Dechlori-nation chemical dosage (Table 7-2)	÷	Strength of dechlori-nation chemical solution (%)

(c7-4)

Dechlori-nation chemical feed rate (mL/min)	= 6 ×	Chlorinated water flow rate (L/sec) (withdrawal or fill)	×	Chlorine residual (mg/L)	×	Dechlori-nation chemical dosage (Table 7-2)	÷	Strength of dechlori-nation chemical solution (%)

These calculators are used in the situation where dechlorination is applied to a continuous flow. The three calculators are used to determine the dechlorination feed rate in either gallons per hour, milliliters per minute with the continuous flow in gallons per minute, or liters per second. The calculators have four factors. Insert the values for each factor and calculate the dechlorination feed rate.

- The first factor is the flow rate of the chlorinated water (withdrawal or fill rate) to be dechlorinated.
- The second factor is the chlorine residual in mg/L.
- The next factor is the dechlorination chemical dosage from Table 7-2.
- The last factor is the strength of the dechlorination chemical solution in %.

Example 1.   The slug method is used for chlorination of 1,000 ft of 10-in. diameter pipeline. A 20% slug volume was selected because the job needed to be completed in about 16 hr. Dechlorinate as the slug exits the pipeline using a 1% sodium sulfite solution. Describe the dechlorination process of the slug as it is slowly withdrawn from the pipeline.

- Use calculator c3-6 and Table 3-6 to determine the volume of the slug.

    1,000 ft × 4.08 gal/ft × 0.2 = 816 gal and the chlorine residual is 100 mg/L

- Use calculator c3-7 to determine the withdrawal rate.

    816 gal ÷ 180 min = 4.5 gal/min

- Use the dechlorination continuous solution feed calculator c7-2.

  *0.006 × 4.5 gal / min × 100 mg / L × 1.96 mg / mg ÷ 1%
  = 5.3 gal / hr feed rate*

Using this method, the slug will last about 3 hr so the total volume of 1% sodium sulfite solution needed is 5.3 gal/hr × 3 hr = 16 gal.

Example 2.   Suppose the slug method was used for 500 m of 150-mm pipeline. A 15% slug was used. The chlorine residual is measured at 80 mg/L. Use 1% ascorbic acid to dechlorinate the water as it is removed from the pipeline. Describe the dechlorination procedure.

- Use calculator c3-6 and Table 3-6 to determine the volume of the slug.

  500 m × 17.66 L / m × 0.15 = 13,245 L and the chlorine residual is 80 mg / L

- Use calculator c3-7 to determine the withdrawal rate.

  *13,245 L / 180 min = 7.4 L / min; convert to L / sec =
  7.4 L / min × 1 min / 60 sec = 0.12 L / sec*

- Use the dechlorination continuous solution feed calculator c7-4.

  *6 × 0.12 L / sec × 2.5 mg / mg × 80 mg / L ÷ 1% =
  144 mL / min feed rate*

Using this method, the slug will last about 3 hr so the total volume of 1% ascorbic acid solution needed is 144 mL/min × 180 min = 25,920 mL or 25 L.

References

APHA, AWWA, and WPF. 1998. *Standard Methods for the Examination of Water and Wastewater.* 20th ed. APHA: Washington, D.C.

AWWA. 2006. *AWWA Manual M20, Chlorine / Chloramine Disinfection.* 2nd ed. AWWA: Denver, Colo.

AwwaRF and AWWA. 2001. *Guidance Manual for Disposal of Chlorinated Water.* Tikkanen, M., Schroeter, J., Leong, L., and Ganesh, R. AWWA: Denver, Colo.

White, G.C. 1999. *Handbook of Chlorination and Alternative Disinfectants.* Wiley-Interscience: New York.

Chapter 8

Pipeline Chlorination Simplified

The previous chapters describe the procedures for pipeline and storage facility chlorination (and dechlorination) that apply to a multitude of situations. This information should be used to correctly and effectively disinfect these water system components. There is an overwhelming array of procedures needed to effectively deal with every possible field situation. Even though all of this information is needed to correctly disinfect pipelines and storage facilities, of all possible sizes and with all the available chemicals under a variety of conditions, there are only a few situations that represent the vast majority of conditions faced by most system operators.

This chapter presents the most common pipeline disinfection procedures needed to address the majority of field conditions. Using these procedures greatly reduces the complexity of disinfection choices and, thus, simplifies field disinfection for most pipelines. Although the procedures listed here represent common practice of most utilities, there are many cases where other choices are appropriate or even required. The information presented throughout this field guide provides the system operator all that is needed to effectively disinfect (and dechlorinate) all types of pipelines and storage facilities.

All of the following steps (from Chapter 3) need to be followed to ensure a safe and efficient pipeline installation. The procedures provided in this chapter are to simplify step #6.

1. Inspection
2. Sanitary construction methods
3. Flushing or cleaning
4. Preventing backflow during installation
5. Providing temporary service
6. Chlorination (chemical disinfection)

7. Final flushing
8. Bacteriological testing
9. Connection to distribution system
10. Documentation

Simplified Pipeline Chlorination Methods

The method selected by most utility system operators to disinfect pipelines generally depends on the size of the pipes. There are three possible methods commonly used for this purpose:

- Tablet (granules)
- Continuous feed
- Slug

Many system operators have abandoned the tablet (granules) method in favor of the other two. A common explanation of this preference is that the granules are often washed to one end of the pipeline during the filling procedure. This often occurs despite using great care to fill the pipeline slowly. When this happens only part of the pipe is exposed to the correct disinfection dosage. Another problem occurs when using tablets because they do not always completely dissolve or they do not dissolve uniformly as the pipe is filled. This results in a similar condition. Tablets and granules are, however, commonly used to disinfect hydrants, valve branches, and other appurtenances.

When employing any of the chlorination methods there are several measurements that are always needed. Some of these are defined by the pipeline measurements and some are the result of the chlorination method selected. The values for the measurements listed in the following chlorination checklist should be obtained.

Using the following procedures provides the values for each of these necessary measurements. Some of the measurements are calculated using the tables and calculator formulas, while some are given as elements of the description of the job.

Chlorination Checklist
❏ The pipeline diameter and length.
❏ Valves, hydrants, tees, and other appurtenances.
❏ Volume of chlorination water to disinfect pipeline (this is the volume for disposal; chlorine residual is needed to determine dechlorination chemical amount needed).
❏ Amount of chlorine chemical needed to complete the disinfection procedure.

Pipelines Less Than 12 in. (300 mm)

The continuous feed method (in Chapter 3) is the most common for pipes less than 6-in. (150-mm) diameter; however, many utilities use this method for 12-in. pipes. The amount of chlorinated water needed for these small pipes is manageable and the method is effective. A 25 mg/L chlorine solution is used to fill the pipeline. After 24 hr, the residual must be at least 10 mg/L. Both calcium hypochlorite and sodium hypochlorite are commonly used to prepare the 25 mg/L chlorination solution. Liquid chlorine is not usually used for this purpose (although it is acceptable) because of safety considerations when handling chlorine cylinders in the field.

Follow the procedures in Chapter 3 for the continuous-feed method. Cleaning the pipeline before beginning the disinfection process is most important. Use Table 8-1 to prepare 25 mg/L feed solution using calcium hypochlorite or sodium hypochlorite. Make sure that the solution is thoroughly mixed prior to feeding into the pipeline. Dechlorinate the discharge, if necessary, according to procedures in Chapter 7.

To use this method, all that is needed is the volume of the pipeline. From this and using the tables and calculators, the amount of chlorine chemical can be determined and, if necessary, the dechlorination requirements can be calculated. The method requires a 25 mg/L dosage, so this is used in the chlorine amount calculator. Follow these steps:

1. Use calculator c8-1, inserting the applicable value from Table 8-2 to calculate the volume of the pipeline and, thus, the volume of 25 mg/L chlorine solution needed to fill the pipeline. Prepare the needed volume (make 25% extra volume to cover spillage, etc.) using Table 8-1.
2. Calculate the amount of chlorine chemical needed using calculator c8-2 and inserting the volume from c8-1 above and the chemical per unit volume shown in Table 8-3.

Table 8-1   Amount of chemical to prepare 25 mg/L chlorine solution for various volumes

gal (L) of 25 mg/L solution	Calcium Hypochlorite g 65%	Sodium Hypochlorite mL 5%	Sodium Hypochlorite mL 10%	Sodium Hypochlorite mL 12.5%
50 (189)	7.3	95	47	38
100 (379)	14.5	189	95	76
150 (567)	21.9	284	141	114
200 (756)	29.2	378	189	152

Only two steps are needed when using these simplified procedures and targeted calculators and tables. Examine the examples below to help understand how to use these tools for this process.

Volume of Chlorination Water Needed to Disinfect the Pipeline

Pipeline volume calculator. To calculate the volume of 25 mg/L chlorine solution needed to fill a pipeline, use this calculator.

(c8-1)

Volume of 25 mg/L solution needed to fill pipeline = Length of pipeline of selected diameter × Volume per unit length (from Table 8-2)

Table 8-2   Pipeline volume for various small diameters

US Units		SI Units	
Pipe Diameter *in.*	*gal/ft*	Pipe Diameter *mm*	*L/m*
2	0.16	50	1.96
2.5	0.25	65	3.32
3	0.37	75	4.42
4	0.65	100	7.85
6	1.47	150	17.66
8	2.61	200	31.40
10	4.08	250	49.06

Amount of chlorine needed calculator.

(c8-2)

$$\text{Amount of chlorine needed} = \text{Volume (from c8-1)} \times \text{Amount of chemical per unit volume (from Table 8-3)}$$

Table 8-3   Amount of chlorine chemical needed per unit volume for 25 mg/L

Calcium Hypochlorite 65%		Sodium Hypochlorite mL 5%		Sodium Hypochlorite mL 10%		Sodium Hypochlorite mL 12.5%	
g/gal	*g/L*	*mL/gal*	*mL/L*	*mL/gal*	*mL/L*	*mL/gal*	*mL/L*
0.146	0.039	1.9	0.503	0.94	0.249	0.76	0.201

For the continuous-feed method, the volume is the total volume of the pipeline previously given. The concentration must be measured but may be assumed to be equivalent to the dosage. These values are needed to calculate the amount of dechlorination chemical needed as shown in Chapter 7.

Example 1.   Calculate the volume of 25 mg/L solution needed to chlorinate a 200-ft long section of 2.5-in. diameter pipeline with 10% sodium hypochlorite.

Use the checklist to ensure all of the information is collected.

❏ The pipeline diameter and length.

2.5-in. diameter and 200 ft long

❏ Valves, hydrants, tees, and other appurtenances.

No valves or hydrants are listed.

❏ Volume of pipeline to be chlorinated.

Using calculator c8-1, volume in gal = 200 ft ×
0.25 gal / ft = 50 gal
This is the volume of chlorinated water for disposal.
Chlorine residual measurement is needed to determine
the amount of dechlorination chemical needed.

❏ Amount of chlorine needed to complete the disinfection procedure.

Using calculator c8-2, mL 10% sodium hypochlorite =
50 gal × 0.94 mL / gal = 47 mL

Example 2.   Calculate the volume of 25 mg/L solution needed to chlorinate a 100-m long section of 75-mm diameter pipeline using 5% sodium hypochlorite.

Use the checklist to ensure all of the information is collected.

❏ The pipeline diameter and length.

75 mm diameter and 100 m long

❏ Valves, hydrants, tees, and other appurtenances.

No valves or hydrants are listed.

❏ Volume of pipeline to be chlorinated.

Using calculator c8-1, volume in L = 100 m ×
4.42 L / m = 442 L

❏ Amount of chlorine needed to complete the disinfection procedure.

Using calculator c8-2, mL 5% sodium hypochlorite =
442 L × 0.503 mL / L = 222 mL

Pipelines 12 in. (300 mm) or Larger

The slug method is preferred for larger pipelines, most commonly larger than 12 in., but many utilities use this method for pipelines 6 in. or larger (150 mm). This method is described in detail in Chapter 3. Generally, a section of pipeline is filled with 100 mg/L chlorine solution, and the slug is moved slowly along the length of the pipeline so that the contact time is at least 3 hr. Valves and hydrants are operated as the slug passes to ensure that they are disinfected. Again, thorough cleaning prior to beginning the disinfection procedure is critical to success.

The slug is moved through the pipeline in one of two ways. It is common to fill a section of pipeline that can be isolated by valves. Many times this is about 200 ft of pipeline. The 100 mg/L chlorine solution is held in this section for at least 3 hr. Then the downstream valves are opened, and water is fed into the upstream end to move the slug to the next section where it is again isolated by valves for at least 3 hr. This process is repeated as necessary to complete disinfection of the entire pipeline length.

A second method for moving the slug is to inject the chlorine into the pipeline while withdrawing water from the end (Figure 8-1). The slug is prepared as needed to ensure at least 3 hr of contact. This slug then continues to move down the pipeline allowing potable water to replace the 100 mg/L chlorine slug as it moves slowly along. The rate of water withdrawal is monitored by a flowmeter. This method is commonly used for very large diameter pipelines because it uses less chlorinated water and, thus, reduces the need for dechlorination and disposal. One negative aspect to this procedure is that it can take many hours because of the slow rate at which the slug must move through the pipeline.

To reduce the options and simplify the process, a 20 percent slug volume (of the total volume of the pipeline to be disinfected) has been selected. This is a convenient amount because the slug fill time will be 3 hr, and the time to move the slug will always be about 15 hr. Therefore, the entire job from set-up to flushing can be completed in just over two shifts. These times are always the same.

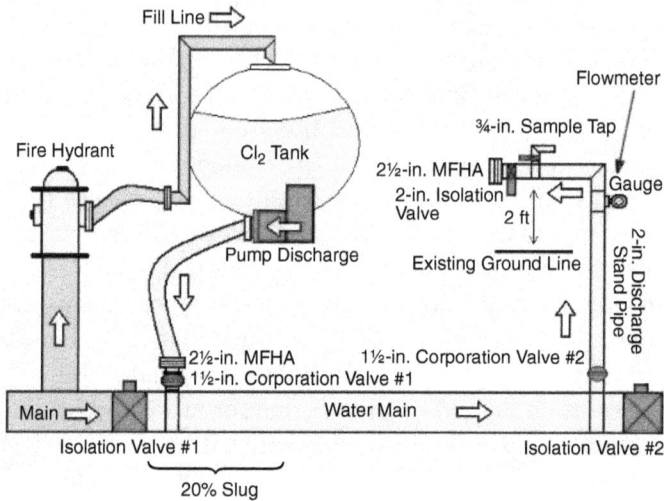

Courtesy Nicole Peschel

Figure 8-1   Pipeline disinfection slug method example

Table 8-4 should be used to calculate the volume of the pipeline and its 20 percent volume of the slug. The 20 percent slug
volume will be the volume of 100 mg/L chlorine solution that will
be needed. The fill rate and the withdrawal rate are the same in
this case so the variables are reduced. The withdrawal rate for a
3-hr contact time is given in Table 8-5. The 100 mg/L chlorine
solution is prepared as needed in a tank or is prepared by matching the fill rate with the feed rate from a 1 percent or 5 percent
chlorine solution source (Table 8-5).

Table 8-4   Pipeline volume for various diameters (also Table 3-6)

US Units		SI Units	
Pipe Diameter *in.*	*gal/ft*	Pipe Diameter *mm*	*L/m*
6	1.47	150	17.66
8	2.61	200	31.40
10	4.08	250	49.06
12	5.87	300	70.65
16	10.44	400	125.60
18	13.21	450	158.96
20	16.31	500	196.25
24	23.49	600	282.60
30	36.70	800	502.40
36	52.85	900	635.85
42	71.93	1,000	785.00
48	93.95	1,200	1,130.40
54	118.90	1,400	1,538.60
60	146.80	1,500	1,766.25
64	167.02	1,600	2,009.60

Table 8-5   Approximate chlorine solution* feed rate at various flow rates for 100 mg/L dosage (from Table 3-11)

US Units			SI Units		
Water Fill Rate *gpm*	Solution Feed Rate		Water Fill Rate *L/sec*	Solution Feed Rate	
	1% Solution *gph*	5% Solution *gph*		1% Solution *mL/min*	5% Solution *mL/min*
1	0.6	0.12	0.1	60	12
2	1.2	0.24	0.2	120	24
3	1.8	0.36	0.3	180	36
4	2.4	0.48	0.4	240	48
5	3	0.6	0.5	300	60
6	3.6	0.72	0.6	360	72
7	4.2	0.84	0.7	420	84
8	4.8	0.96	0.8	480	96

Table continued next page.

Table 8-5   Approximate chlorine solution* feed rate at various flow rates for 100 mg/L dosage (from Table 3-11) *(continued)*

US Units			SI Units		
Water Fill Rate	Solution Feed Rate		Water Fill Rate	Solution Feed Rate	
	1% Solution	5% Solution		1% Solution	5% Solution
gpm	*gph*	*gph*	*L/sec*	*mL/min*	*mLlmin*
9	5.4	1.08	0.9	540	108
10	6	1.2	1	600	120
15	9	1.8	1.1	660	132
20	12	2.4	1.2	720	144
25	15	3	1.3	780	156
30	18	3.6	1.4	840	168
35	21	4.2	1.5	900	180
40	24	4.8	1.6	960	192
45	27	5.4	1.7	1,020	204
50	30	6	1.8	1,080	216
55	33	6.6	1.9	1,140	228
60	36	7.2	2	1,200	240
65	39	7.8	2.5	1,500	300
70	42	8.4	3	1,800	360
75	45	9	3.5	2,100	420
80	48	9.6	4	2,400	480
85	51	10.2	4.5	2,700	540
90	54	10.8	5	3,000	600
95	57	11.4	5.5	3,300	660
100	60	12	6	3,600	720
110	66	13.2	6.5	3,900	780
120	72	14.4	7	4,200	840
130	78	15.6	7.5	4,500	900
140	84	16.8	8	4,800	960
150	90	18	8.5	5,100	1,020
160	96	19.2	9	5,400	1,080
170	102	20.4	9.5	5,700	1,140
180	108	21.6	10	6,000	1,200
190	114	22.8			
200	120	24			

* Chlorine solution is prepared by mixing 1 lb calcium hypochlorite (65%) with 8 gal of water or diluting 10% sodium hypochlorite 1 gal with 9 gal of water. Five percent chlorine solution is prepared by mixing 5 lb calcium hypochlorite (65%) with 8 gal of water or dilution 10% sodium hypochlorite 1 gal with 1 gal of water.

Simplified Slug Method for Larger Pipelines

The simplified method assumes a slug that is 20% of the total length of the pipeline to be disinfected. Also, the rate that the slug is moved along the pipeline is such that the contact time is 3 hr at every point. Limiting the procedure to these parameters reduces the calculations to only three steps.

1. Calculator c8-3 is used to calculate the volume of the 20% slug. The length of the pipeline and the appropriate value from Table 8-3 are used in this calculation.

2. Next, calculator c8-4 is used to determine the withdrawal rate (the rate that water is moved along the pipeline to achieve the necessary 3-hr contact time). The volume of the 20% slug calculated in step 1 is divided by the 180 minutes (3 hr). The withdrawal rate (same as fill rate) is also used to find the matching chlorine solution feed rate for the 1% or 5% strength solution.

3. The volume of 1% chlorine solution needed to prepare the 20% slug with a dosage of 100 mg/L is calculated using c8-5. This is the amount of 1% solution that needed to be on hand (add about 25% extra for contingencies) to complete the procedure.

Limiting the variables and using this three-step process greatly simplifies a process that can become confusing and complex. Examine the examples below to gain further insight regarding how to use this procedure.

Volume of Chlorination Water to Disinfect the Pipeline

Volume of 20 percent slug calculator.

(c8-3)

Volume of 20% slug	=	Total length of pipe to be disinfected	×	Volume/ft or volume/m (from Table 8-3)	×	0.2

For the 20 percent slug method, calculate the withdrawal rate using the following calculator to ensure 3-hr (180-min) contact time at all points along the pipeline. Choose the matching chlorination chemical solution feed rate from Table 8-5.

Withdrawal rate calculator.

(c8-4)

Withdrawal rate or fill rate for 3 hr contact (gpm or L/min) = Volume of 20% slug (from c8-3) ÷ 180 min

Amount of Chlorine Needed to Complete the Disinfection Procedure

Chlorination chemical amount calculator.

(c8-5)

Volume of 1% chlorine solution needed for 20% slug = Volume of 20% slug (from c8-3) × 0.01

Example 1.   Create a 20% slug of 100 mg/L chlorine concentration for a 10-in. diameter pipeline. Disinfect 1,500 ft of new pipeline. What is the water withdrawal rate that will result in at least a 3-hr contact time along the pipeline? If 1% sodium hypochlorite is injected to form the slug, what is the feed rate? How much 10% sodium hypochlorite will be needed to complete the job?

Use the checklist to ensure all of the information is collected.
- ❏ The pipeline diameter and length.
    10-in. diameter and 1,500-ft length
- ❏ Valves, hydrants, tees, and other appurtenances.
    No valves or hydrants are listed.
- ❏ Volume of chlorination water to disinfect pipeline
    Using calculator c8-3, volume = 1,500 ft × 4.08 gal / ft × 0.2 = 1,224 gal

☐ Amount of chlorine needed to complete the disinfection procedure.

> *Using calculator c8-5, 1% sodium hypochlorite =*
> *1,224 gal × 0.01 = 12.24 gal*
> *1 gal of 10% sodium hypochlorite is needed to make*
> *10 gal of 1% so 1.2 gal of 10% is needed to make*
> *12.24 gal of 1% solution*

☞ The withdrawal rate for a 20% slug (c8-4) = 1,224 gal ÷ 180 min (3 hr) = 6.8 gal/min

☞ 1% sodium hypochlorite feed rate from Table 8-5 is 4.2 g/hr for 7 gpm fill rate. This will result in a chlorine dosage slightly higher than 100 mg/L.

Example 2.   Create a 20% slug of 100 mg/L chlorine concentration for a 500-mm diameter pipeline. Disinfect 1,000 m of new pipeline. What is the water withdrawal rate that will result in at least a 3-hr contact time along the pipeline? If 1% hypochlorite solution (prepared from calcium hypochlorite) is injected to form the slug, what is the feed rate?

Use the checklist to ensure all of the information is collected.

☐ The pipeline diameter and length.

> *500-mm diameter and 1,000-m length*

☐ Valves, hydrants, tees, and other appurtenances.

> *No valves or hydrants are listed.*

☐ Volume of chlorination water to disinfect pipeline

> *Using calculator c8-3, volume = 1,000 m × 196.25 L / m*
> *× 0.2 = 39,250 L*

☐ Amount of chlorine needed to complete the disinfection procedure.

> *Using calculator c8-5, 1% hypochlorite = 39,250 L ×*
> *0.01 = 390 L*
> *1 lb of 65% calcium hypochlorite is needed to make*
> *8 gal of 1% solution, so 454 g or 0.454 kg is needed*
> *for (3.785 × 8) 30.28 L of 1% solution. Therefore,*
> *0.454 kg / 30.28L × 390 L = 5.8 kg*

☞ The withdrawal rate for a 20% slug (c8-4) = 39,250 L × 180 min (3 hr) = 218 L/min.

Convert to L/sec: 218 L/min × 60 sec/min = 3.6 L/sec

☞ 1% hypochlorite feed rate from Table 8-5 is 2,400 mL/min for a fill rate of 4 L/sec. This will result in a chlorine dosage slightly higher than 100 mg/L.

Appendix A

Calculators

Below are all of the calculators presented in this field guide. They are summarized here for convenience. The detailed derivation of the formulas is shown as well as the calculator as it appears in the body of the field guide. Note that the calculators in the field guide are stripped of the details and only show a single constant that may be the combination of several. Refer to the chapters and pages indicated for detailed instructions and examples.

Pipeline Continuous-Feed Method Calculator

To calculate the chlorine feed rate to achieve 25 mg/L inside a pipeline (does not account for chlorine demand).

gal/min × 25 mg/L × 3.785 L/gal × g/1,000 mg × lb/454 g × 60 min/hr × 24 hr/day = lb/day

(c3-1)

Liquid chlorine feed rate (lb/day)	=	Water fill rate (gpm) total	×	0.3

L/sec × 25 mg/L × g/1,000 mg × 60 sec/min × 60 min/hr = g/hr

(c3-2)

Liquid chlorine feed rate (g/hr)	=	Water fill rate (L/sec) total	×	90

$$\text{gal/min} \times 25 \text{ mg/L} \times 3.785 \text{ L/gal} \div \% \times 0.01 \times 1{,}000 \text{ mg/g} \times$$
$$1 \text{ mL/g} = \text{mL/min}$$

(c3-3)

Sodium hypochlorite solution feed rate (mL/min) = 9.46 × Water fill rate (gpm) total ÷ Strength of sodium hypochlorite feed solution (%)

(c3-4)

Sodium hypochlorite solution feed rate (gal/hr) = 0.15 × Water fill rate (gpm) total ÷ Strength of sodium hypochlorite feed solution (%)

$$\text{L/sec} \times 25 \text{ mg/L} \times 1{,}000 \text{ mL/L} \times 60 \text{ sec/min} \times 0.001 \text{ L/mg} \times$$
$$0.01 \div \% = \text{mL/min}$$

(c3-5)

Sodium hypochlorite solution feed rate (mL/min) = 150 × Water fill rate (L/sec) total ÷ Strength of sodium hypochlorite feed solution (%)

Tank Volume Calculators

Rectangular or Square Tanks

(c5-1)

Length, in ft × Width, in ft × Maximum depth of water, in ft × 7.48 gal/ft^3 = Volume, in gal

(c5-2)

Length, in m × Width, in m × Maximum depth of water, in m = Volume, in m^3

Vertical (upright) Cylindrical Tanks

$$3.14 \times \text{dia. ft} \times \text{dia. ft} \times \text{height ft} \times 7.48 \text{ gal/ft}^3 \div 4 = \text{gal}$$

$$(c5\text{-}3)$$

$$5.87 \times \boxed{\begin{array}{c}\text{Diameter,}\\\text{in ft}\end{array}} \times \boxed{\begin{array}{c}\text{Diameter,}\\\text{in ft}\end{array}} \times \boxed{\begin{array}{c}\text{Maximum}\\\text{depth of}\\\text{water, in ft}\end{array}} = \begin{array}{c}\text{Volume,}\\\text{in gal}\end{array}$$

$$3.14 \times \text{dia. m} \times \text{dia. m} \times \text{height m} \div 4 = \text{m}^3$$

$$(c5\text{-}4)$$

$$0.785 \times \boxed{\begin{array}{c}\text{Diameter,}\\\text{in m}\end{array}} \times \boxed{\begin{array}{c}\text{Diameter,}\\\text{in m}\end{array}} \times \boxed{\begin{array}{c}\text{Maximum}\\\text{depth of}\\\text{water, in m}\end{array}} = \begin{array}{c}\text{Volume,}\\\text{in m}^3\end{array}$$

Full Storage Facility Chlorination Method Chemical Calculator

The approximate amount of chlorination chemical needed for 10 mg/L chlorine concentration for a given size tank is calculated as follows (use either US or SI units):

$$(c5\text{-}5)$$

1	2	3	4	
Volume (from tank shape equations)	× of chemical (from Table 5-2)	× strength factor = strength in table/strength of chemical selected	÷ (from Table 5-2)	= of chemical needed

(Column headers: 1 — Volume (from tank shape equations); 2 — Amount of chemical (from Table 5-2); 3 — Chemical strength factor = strength in table/strength of chemical selected; 4 — Volume (from Table 5-2); = Amount of chemical needed)

Storage Facility Chlorinate and Fill Method Calculator

The approximate amount of chlorination chemical needed for 50 mg/L chlorine concentration for a given size tank, 5% volume, is calculated as follows (use either US or SI units):

(c5-6)

1		2		3		4		
Volume (from tank shape equations × 0.05)	×	Amount of chemical (from Table 5-5)	×	Chemical strength factor = strength in table/strength of chemical selected	÷	Volume (from Table 5-5; 100,000 for US units and 100 for SI units)	=	Amount of chemical needed

Dechlorination Chemical Calculator

(c7-1)

1		2		3		4		
Volume of water to be dechlorinated (from Tables 3-5 and 5-1, and calculators 5-1, 5-2, 5-3, and 5-4)	×	Amount of chemical (from Table 7-5)	×	Chlorine residual, mg/L	÷	Volume (from Table 7-5; either 100,000 gal or 100 m^3)	=	Total amount of chemical needed

Dechlorination Continuous Solution Feed Chemical Calculator

gal/hr = gal/min × mgC/L × mgD/mgC × 60 min/hr ÷ (% strength × 10,000 mg/L/%)*

*mgC is the milligrams of chlorine and mgD is the milligrams of dechlorination chemical

(c7-2)

Dechlorination chemical feed rate (gph)		Chlorinated water flow rate (gpm) (withdrawal or fill)		Chlorine residual (mg/L)		Dechlorination chemical dosage (Table 7-2)		Strength of dechlorination chemical solution (%)
= 0.006 ×			×		×		÷	

$$\text{mL/min} = \text{gal/hr (from c7-2)} \times 3{,}785 \text{ mL/gal} \times \text{hr/60 min}$$

$$(c7\text{-}3)$$

Dechlorination chemical feed rate (mL/min)	= 0.038 ×	Chlorinated water flow rate (gpm) (withdrawal or fill)	×	Chlorine residual (mg/L)	×	Dechlorination chemical dosage (Table 7-2)	÷	Strength of dechlorination chemical solution (%)

$$\text{mL/min} = \text{L/sec} \times \text{mgC/L} \times \text{mgD/mgC} \times 60\text{sec/min} \times$$
$$1{,}000 \text{ mL/L} \div (\% \text{ strength} \times 10{,}000 \text{ mg/L/\%})$$

*mgC is the milligrams of chlorine and mgD is the milligrams of dechlorination chemical

$$(c7\text{-}4)$$

Dechlorination chemical feed rate (mL/min)	= 6 ×	Chlorinated water flow rate (L/sec) (withdrawal or fill)	×	Chlorine residual (mg/L)	×	Dechlorination chemical dosage (Table 7-2)	÷	Strength of dechlorination chemical solution (%)

Pipeline Volume Calculator—Continuous-Feed Method

To calculate the volume of 25 mg/L chlorine solution needed to fill a pipeline, use this calculator.

$$(c8\text{-}1)$$

Volume of 25 mg/L solution needed to fill pipeline	=	Length of pipeline of selected diameter	×	Volume per unit length (from Table 8-2)

Chlorine Calculator

$$(c8\text{-}2)$$

Amount of chlorine needed	=	Volume (from c8-1)	×	Amount of chemical per unit volume (from Table 8-3)

20 Percent Slug Method Withdrawal Rate Calculator

(c8-3)

| Volume of 20% slug | = | Total length of pipe to be disinfected | × | Volume/ft or volume/m (from Table 8-3) | × | 0.2 |

(c8-4)

| Withdrawal rate or fill rate for 3 hr contact (gpm or L/min) | = | Volume of 20% slug (from c8-3) | ÷ | 180 min |

Chlorination Chemical Amount Calculator

(c8-5)

| Volume of 1% chlorine solution needed for 20% slug | = | Volume of 20% slug (from c8-3) | × | 0.01 |

Index

Note: *f.* indicates figure; *t.* indicates table.

A

B

C

D

F

L

Liquid chlorine, 9–10
  and lower pH, 8–9
  reaction with water, 7
  safety considerations, 9–10

P

Pathogens present in soil or external water, 2–3, *3f.*
pH, effect on chlorine species, 7, *8f.*
Pipeline repairs, 59
  bacteriological testing, 64
  chlorination of open trenches, 60–61
  chlorination of pipe interior and fittings, 61, 62
  chlorination of repaired section, 61–62, *63t.*
  chlorination procedures, 60–63
  condition and storage of repair parts, 59–60
  documentation, 64
  excavation dewatering, 60, *61f.*
  flushing, 62
  sanitary practices, 60
  tapping sleeves, 62–63
Pipelines
  approximate volume for various diameters, *29t.*
  keeping pipe clean and dry, 14, *15f.*
  temporary pipeline materials, 26
  volume calculator, 112
  volume for various larger diameters, *117t.*
  volume for various small diameters, *113t.*
  See also Flushing and cleaning, Sanitary construction methods
Plumbing codes, 4

S

Sanitary construction methods, 13–14
  and flooding, 16
  joints, 14

W

Waterborne-disease outbreaks
  due to distribution system deficiencies, 2, 2*f.*
  1971–2002, 1, 1*f.*, 2*f.*